36th Congress, 2d Session. } HOUSE OF REPRESENTATIVES. { Ex. Doc. No. 44.

MILITARY ROAD FROM FORT BENTON TO FORT WALLA-WALLA.

LETTER

FROM

THE SECRETARY OF WAR,

TRANSMITTING

The report of Lieutenant Mullan, in charge of the construction of the military road from Fort Benton to Fort Walla-Walla.

January 25, 1861.—Referred to the Committee on Naval Affairs, and ordered to be printed.

War Department, *January* 23, 1861.

Sir: I have the honor to transmit herewith a communication from the officer in charge of the Office of Explorations and Surveys, accompanied by a copy of the paper called for in the resolution of the House of Representatives of the 18th ultimo, by which the Secretary of War is requested "to communicate to the House the reports of Lieutenant John Mullan, United States army, of his operations in charge of the wagon road from Fort Benton to the Walla-Walla."

Very respectfully, your obedient servant,

J. HOLT, *Secretary of War.*

Hon. William Pennington,
Speaker of the House of Representatives.

War Department, Office Explorations and Surveys,
Washington, January 22, 1861.

Sir: I have the honor to submit herewith the reports of Lieutenant John Mullan, United States army, upon his operations in charge of the construction of the military road from Fort Benton to Fort Walla-Walla, called for by a resolution of the House of Representatives of December 18, 1860. The reports are numbered from I to XVII; the sub-reports, from assistants, are marked from A to T.

Very respectfully, your obedient servant,

A. A. HUMPHREYS,
Captain Top'l Engineers, in charge.

Hon. J. Holt, *Secretary of War.*

I.

CAMP ON SNAKE RIVER, W. T.,
Opposite mouth of Toukanon, July 3, 1859.

SIR: I have delayed until the present date and place to give you the details concerning my movements, in order that I might be enabled to let you know that we were well underway.

Arriving at the Dalles on May 10, I organized by 25th, moving into a temporary camp on the Five and Ten Miles creek, where we remained until June 8, when I took up a line of March for Fort Walla-Walla, reaching that point on the 22d instant, where I tarried for a few days for repairs, &c.

Here my escort joined me, it having gone forward, *via* the Columbia river, to Fort Walla-Walla. This escort is composed of 100 men, 3d artillery, under Lieutenant White, commanding, Lieutenants Howard and Lyon.

We moved from Walla-Walla on the morning of the 25th of June. I would state that on our way to Walla-Walla from the Dalles my party worked and improved the road from day to day, facilitating thus not only our own progress, but so materially improved the road—which, in consequence of high water, had become much damaged at many points—that future travel in the interior must reap its benefits. From Fort Walla northward, both on account of wood, water, and grass, and the general character of the country, I was led to locate the road to the mouth of the Toukanon, (Fort Taylor of last year,) 3 miles east of the mouth of the Pelouse. Private interests have been at work to induce me to locate more to the west, in order that ferries, &c., now established on the river, might be availed of, but looking to an eastern connexion and the favorable character of the country, and excellent wood, grass, and water; and at short intervals this location has *every* advantage, whereas the one by the Pelouse crossing involves a 26 miles' march without water or wood.

Moving, then, on the 25th for the mouth of the Toukanon, our work began. We crossed the dry creek, repairing here a bridge and establishing a camp; marked it by a post buried three feet in the ground, and marked this M. R., (military road,) 9 miles.

Moving still northward across the Touchet, improving and working the road from point to point, we encamped upon its banks for two days, and threw over it a strong, substantial bridge, built of the only timber at hand, viz: the cottonwood. This stream is difficult at high water. But selecting a suitable spot we constructed a bridge with three spans that is five feet above the high water mark. Moving still northward, we worked the road along the *Reed creek,* encamping at its head.

Moving still northward with an advanced working party, we improved and worked the road along the Toukanon to its mouth, crossing it three miles from its mouth. The distance then from old Fort Walla-Walla to this point is 78½ miles, and from the new fort 49 miles, reaching this point on the 28th ultimo. I have been much delayed and annoyed at this point in crossing the Snake river. The river is frightfully swollen in consequence of the great freshet of this year; and it has been with the most extreme difficulty that we to-day effected the crossing of our last load. The animals feared taking the water, and had to swim by being led across, involving much time, labor, and risk; but to-day I am happy to state we completed the crossing, involving five days. We move at daybreak in the morning still northward for the Pelouse; 14 miles from its mouth hence I will follow it for, say, eight or ten miles, when I think the valley of the Schmokah creek will afford me the best location; my probable course will then be up its valley to a convenient and suitable point to turn northeasterly for the Pyramid Butte, whence my line will be for the southern edge of the Cœur d'Alène lake, whence, or earlier, I will again write you, giving you the results of our progress and our then next section of our line.

My advanced party has been ahead for the last three days working a section of five miles of our line hence to the Pelouse. It is now finished, and will offer no difficulty to our progress for the present.

This involved some difficulty owing to the want of water, but by packing the water to the working parties all has gone well.

I regret to report the loss of one of Lieutenant White's men by drowning while crossing the Snake river. He had been above the camp for wood, which he floated down the stream in the form of a raft; being upon which, the current, which is at the rate of five or six miles per hour, carried him so far below that, jumping off it, it was impossible to save him.

With the exception of the loss of some of our animals by drowning, all else has gone well.

While organizing at the Dalles, in order that my party should be employed as industriously as possible, and in order that an equivalent might be returned by the different members of the party, I sent small parties as follows:

A party under Mr. Kollecki to Fort Walla-Walla, with all the astronomical instruments, chronometers, &c., where they remained a month; established an observatory by the assistance of Captain Kirkham, who has uniformly co-operated with me.

Here they had excellent weather, and, putting up their transit, have obtained not less than from three to four hundred sets of data, the exact determinations of this point.

Thus is fixed a good initial point. The different sets for longitude are not yet worked up, and only for latitude approximately; but as leisure is afforded, they will be put under way and completed in the field if possible. In addition to gaining this desideratum, much attention was given to the meteorology of the place, the different mountain peaks measured, and much topographical information gained. I express my thanks to Mr. Kollecki and Mr. Weisner, under whose general charge this has been done.

A small party under Mr. Conway R. Howard moved from the Dalles, taking the trail *via* the Columbia river, and have been very successful in running a line of level to old Walla-Walla But on reaching that point, in adjusting their level one of the screws became broken, and it has involved sending it to the Dalles for repairs, thus making a delay of some ten days. This party has done much work ; made a map of the line of the river and also along the river, and has furnished the detailed data of cost of locating a rail or tramway along the Columbia. They will continue their work to the mouth of the Snake river, thence to the mouth of the Pelouse, when, their work ending, they will join the main working party in advance.

I regret to report the sickness of Mr. Howard, however, and his necessity to return to Fort Walla-Walla, where he is now under the medical care of Dr. McParlin at the post. A swelling in his limbs, due from cold, doubtless, has induced his present indisposition. The charge of the party devolved upon Mr. De Lacey, the civil engineer lately in the employ of Lieutenant Mendell, and who now has charge of the party. Connected with this party is a Mr. George H. Smitt, who joined me at the Dalles with a letter of appointment in my expedition from the Secretary of War, bearing date of April 8, 1859, employed as civil engineer.

The movement of the command under Major Lugenbeel for Fort Colville was availed of, and Mr. Engle, at the Major's request, moved with him, making an excellent map of the line from this point to Fort Colville, which connects with our work of last year.

Mr. Sohon, who was with me last year, and who is again in my employ, moved from the Dalles, and has gone forward in advance to examine the road towards the head of the Pelouse, reported by the Indians as practicable for a wagon road thence to the Bitter Root valley. He moved with one man and an Indian, and reached a point two hundred miles eastward ; but the Indians have not only thrown every obstacle in his way, but his life has become much endangered. I received an express from him last night, giving the detail of his progress ; and an express goes to him to-morrow morning, with instructions to rejoin my party, which takes up its line of march for the Cœur d'Alène Mission.

From representations made by parties I was led to suppose that a shorter and more practicable route lay by the upper waters of the Pelouse eastward, and determined to investigate it, and Mr. Sohon started with instructions to this effect. Found one hundred and seventy-five miles good, when the mountains became rugged and difficult, thick timber and underbrush, and the route generally unfavorable. The disposition of the Indians, too, is of that character as will not do to trust, especially with small parties.

Under these circumstances, the location will still be *via* the Cœur d'Alène Mission. The Indians are still ill-disposed towards the location of the road through their lands ; and though they may commit no overt act or attack, still I believe they will do all they can to annoy us by running off animals; and every vigilance shall be given, so that we are not taken unawares. Of course, with the party scattered along the line, the temptation to them is great. If they commit any overt act, and do not make the most ample amends, we shall give them a good thrashing, which, I find, is the best lesson to teach them. Be assured that our progress shall not again be disturbed by them, if it can be avoided.

Our work again commenced from Fort Dalles, with a view of correcting any errors that might have been committed last year, which has been continued to this point.

Mr. Engle, with one man, left this point yesterday to go up the Snake river to the mouth of the Clearwater, and thence northward to Tatuna, but will rejoin me at the Pyramid Butte. An accurate survey of this river I deem now to be of the utmost importance, in view of the complete success by steam navigation up the Columbia and Snake river to points already demanded, and as far as demanded by the demands of trade and travel. I am sanguine then steamers can ascend the Snake river to the mouth of the Clearwater. They have already ascended it to this point.

I have thus given you a general exhibit of what has been done since our arrival, and can state that by to-morrow morning we shall be well under way for the mountains.

I have been compelled to organize a larger party than at first contemplated, but I do not deem that I have one man too many. The work to be done is heavy, and the means must be proportionate.

I shall, however, towards the autumn, materially diminish my force, sending a part home *via* the Missouri, and a part *via* Fort Walla-Walla to the settlements. But I deemed, as the heaviest work would be within the first three months, that true economy dictated the supplying a larger force during that time.

My accounts for the quarter ending 30th June will be ready in a few days, and be sent to the department. As the expenses were incurred only towards the end of the quarter, only the quarterly papers are forwarded. I wrote twice to say that, as the appropriation of $100,000 was now available, that I would respectfully request that $60,000 of the amount be placed to my credit in New York, and $40,000 in San Francisco; and, supposing that my requests would be acceded to, I have been compelled, in organizing my party for purchases, &c., to draw on New York, thinking that by the time my drafts reached the east, which would be by the 1st of July, that you would have placed to my credit the necessary amount as asked for; and should this not have been done, of course my drafts have been protested. Again, unless it has been already done, would I ask that sixty thousand dollars ($60,000) be placed to my credit in New York, and the balance, $40,000, in San Francisco. This is absolutely necessary for my operations.

The escort, with its employés, now number about one hundred and forty men; my party, all told, some ninety men—a total of two hundred and thirty men—a force sufficiently strong to repel any attack of Indians. Rumors have reached us to the effect that the Blackfeet intend opposing our progress through their country. I give it for what it is worth; but shall abide our time to report the details.

Be assured, captain, that unless some untoward circumstances intervene to prevent, we shall be across the Bitter Root mountains by the middle of October.

The Indians, thus far, are all friendly, and I fear nothing from them.

Trusting that my course, thus far, will meet with the approbation of the department, and requesting that all communications may be directed to Fort Walla-Walla, Washington Territory, I am, most truly, your obedient servant,

JOHN MULLAN,
Lieutenant United States Army,
In charge of Military Road from Walla-Walla to Fort Benton.

Captain A. A. HUMPHREYS,
U. S. Top. Engineers, in charge of Office Exp. and Surveys.

II.

FORT WALLA-WALLA AND FORT BENTON MILITARY ROAD EXPEDITION.

CAMP ON ST. JOSEPH'S RIVER,
July 31, 1859.

SIR: My communication of 3d of July gave you the results of our progress and general movements up to Snake river, which stream having crossed we moved forward on the morning of the 4th, our road having been completed for some fourteen miles to the Pelouse. Here it became necessary to keep a small exploring party in advance, selecting and choosing the best line of location. From the general topographical character of the country, we saw that our line must lay up the valley of the Pelouse to its highest practicable point, or follow the line of some of its northern tributaries. Mr. Sohon, who had examined the valley of Schmokah, found it practicable, with water in pools, but it required that the work should follow along side hills, involving much work; and, fearing that these pools of water did not last throughout the whole season, I preferred following the valleys of running streams, and hence followed along the main Pelouse to the mouth of the Mocahlissiah. Here, finding that the Pelouse bent too far northward, and the Mocahlissiah lending its position to the proposed line of our road, I determined to examine it. Sending forward Mr. Kollecki with two men, he followed it for sixty miles, to and beyond the Pyramid Butte. He found it practicable for a wagon road for some twenty or thirty miles; but from that distance eastward the stream flowed through a rocky cañon, rendering its line unfavorable and impracticable. He ascended the Pyramid Butte, from which he had a fine view of the country for many miles in all directions, when it became evident that our best if not only line was to follow the high table land that lay between the valley of the Mocahlissiah and its more northern tributaries. His trip resulted in the location of the Pyramid Butte, a general knowledge of the topography of the country, and the fact that any line by the Mocahlissiah, by many called the main Pelouse, was impracticable. From the point where we first struck the Pelouse to the mouth of the Mocahlissiah, a distance of nineteen miles, involved much work, at one point rock blasting for three days. We reached the mouth of the Mocahlissiah on the 10th of July. Here, encamping in good grass, and among the first pine timber met with on the road, we started our exploring party in advance, having determined to follow a line as direct as possible for the southern edge of the Cœur d'Alène lake, governed, of course, by camping ground along the route.

Mr. Sohon and a friendry Indian who had joined us, moving northward, found towards the headwaters of the Oray-tay-ouse, a beautiful table land, which afforded us, with work, an excellent location. Continuing thence along the proposed line, we crossed to Teto-Teto-u-supe, the Aqua-ayo-nepe, the Tril-Tril-po-vetz-sin, the Nedlyhuald creeks, and reached our heaviest work in the timber on the evening of 16th July. Along this line we had much work, but with a fatigue party from our escort of twenty-one men we were enabled to proceed, making good time. Having now (16th July) reached the timber, our severe work commenced; and it became necessary to select the most favorable camp for grazing our animals, and so arranging the command that our work should progress, and at the same time protection be given our different parties in camp and at work. We selected a small valley at the edge of the timber, some six miles west of the mouth of St. Joseph's river, where we remained in camp until the 23d of July. During this time the command made the road through the six miles of timber to the edge of a spur making down between the outlets of the Pound lake and the mouth of the St. Joseph's river. This spur involved some heavy work in rock. They also built a strong, substantial bridge of pine over the outlet of Pound lake, sixty feet broad, with one pier and two heavy abutments, and some three feet above the highest water mark. Our road thence, for two miles, lay along the left bank of the St. Joseph's river, through a strip of cottonwood, which was cleared and made practicable by the fatigue party, under general charge of Mr. Sohon, my guide and interpreter.

Finding here good grass, and a generally favorable location, we moved the whole command over, where we encamped on the 24th July, (Sunday.)

Moving forward with a small advanced party, I found the best point for crossing the river would be to follow the valley on its left bank for three miles still further to the east, over a fine prairie bottom, and directly opposite to the spur of the divide that would lead us to the Cœur d'Alène river. To this point our command moved in the afternoon of the 24th of July, and from which point I now write you. Here the St. Joseph's is two hundred and forty feet broad and thirty-seven feet deep; its usual depth being thirty-five, and with no current. Finding the timber not as convenient as desired, and all other contingencies considered, I determined that here, and also for the Cœur d'Alène river, the same size as this, that a large flatboat would not only be the best, but, indeed, the most suitable, for one or other purposes; and I regarded that any delay here involved would be of no loss to whatsoever command or party would be here passing. For the valley of the St. Joseph's is the best grassed (it and the Cœur d'Alène being of the same character) valley that we found in the Cœur d'Alène mountains, and any party moving eastward or westward would halt here to avail themselves of its fine grazing, and, in case of high freshets, would be less liable to be carried away and destroyed than any bridge erected. Besides, a bridge of a proper character would involve nearly a whole season over these two streams. Having concluded, therefore, that a large flat on this sluggish stream was best suited to our purposes, I set the men to work in the timber with whip-saws, and, after three days' work, have procured timber and plank enough for a boat forty-two feet long and twelve feet broad, which will be completed to-morrow, when we shall begin the crossing of the St. Joseph's. The men are also now engaged in getting out the timber and planks for the boat for the Cœur d'Alène river. As soon as the command shall have crossed the stream, I shall send the boat already completed around by the Cœur d'Alène lake to the Cœur d'Alène river, and thence to the point of crossing, and the one now in progress of construction will be left here.

Finding fine pitch pine in the mountains, we were enabled to burn pitch enough for our boats. While this has been done, I move with a small party in advance to explore the best line to the Cœur d'Alène river, as the line followed by the Indian trail leads over impassable spurs that would, if practicable at all, involve a season's labor. This exploration involved four days in the timber and thick underbrush.

Finding a practicable location that led through a wooded cañon for four miles, I moved a working party of twenty men to Cœur d'Alène river, to work from the mouth of a ravine southward, and have a party of thirty men working from the head of the cañon, or ravine, northward ; and I trust by 3d August that the whole line will be opened, and by the morning of the 4th or 5th of August I shall move the whole command over to the valley of the Cœur d'Alène river ; and, if all things should favor me, I am now in hopes of reaching the Cœur d'Alène Mission by 8th of August at farthest, from which point I shall again write you.

From the time we first struck the timber to the Cœur d'Alène Mission, the work has been more severe than I had any idea of, but it has not as yet discouraged us, but nerved us to the greater exertion ; and I trust that the work ahead will not be so severe as to prevent us reaching the Bitter Root in good season.

Here the fallen timber in the Cœur d'Alène mountains has been lying for years, piled up at such heights that the Indians, to avoid it, have made the most tortuous trail imaginable, up and down hill, in order to avoid which, or rather to cut which, we have much hard and tedious work, rendering our progress slow ; but when once made, our route will be an excellent one. We are now 199 miles from old Fort Walla-Walla and 169 from the new fort; and I regard the route followed as one of the best lines that the country affords, with good grass, wood, and water along the whole line, affording camping ground at short intervals. We have marked the route by mile-posts along the whole line, each branded "M. R., — miles." With regard to these I will speak hereafter. The miles thereon marked is the distance from Walla-Walla.

While the road has been worked, and our camp made for many days at a time at one place, favorable opportunities have occurred for reconnoitring and mapping the country for many miles around. These have been all availed of, and, since I last wrote you, the following side work has been done. Mr. Engle, when I last wrote you, was absent examining the Snake river to the mouth of the Clearwater or Red Wolf's crossing. From this point we moved northward to Tatuna Hill, connected with Mr. Sohon's work, and joined my party at the mouth of the Schmokah creek. Mr. Kollecki fixed the position of Pyramid Butte, and then mapped the valley of the Mocahlissiah, and since, availing ourselves of the barge of the Jesuit priests from the Mission, he, with a small party, went around the Cœur d'Alène lake, making an excellent map of it, and yesterday joined my party at this point. He found the lake embosomed in a rugged bend of mountains, some seventy miles around, and, in places, one hundred feet deep, and in other places could find no bottom. There is no low land around the lake whatever. Nothing but high, rugged, pine-clad mountains girt it on every side. He, with Mr. Weisner, will go to-morrow by water with the barge to the Cœur d'Alène Mission, to there put up their transit and prepare for the approaching culminations they will be enabled to observe there for once a month, as our progress eastward from that point must be slow and tedious. We are now some ten miles below the forks of the St. Joseph's, and, in order to ascertain the true character of the country near its headwaters, I have directed Mr. Engle and Mr. Sohon to proceed early to-morrow morning, exploring both forks, and to join me on the Cœur d'Alène river. The party under Mr. Howard has not yet joined me, but I expect them every day. When I last heard from Mr. Howard (July 11) he was still sick in bed at Fort Walla-Walla, with a swelling in his groin.

The accident to our level has been repaired and the party resumed work, and by this time must have finished, and will soon join me. It has taken them longer than anticipated, but their accidents have been unavoidable. This will give you, therefore, a general exhibit of all the work done, or now in progress up to date, with the exception of Mr. Sohon's examination from the Tatuna Hill north and eastward.

As I informed you in my last communication, thinking that the road spoken of by the Indians, if practicable, would be our best line for the Bitter Root valley, I moved Mr. Sohon forward to its examination. He reached the camas plains of the Cœur d'Alène on the 19th of June. Here he was to obtain a guide. Instead of finding the Indian true to his word, he was absent afar in the mountains. Another Indian, pretending to know somewhat of the route, proposed to guide him. Taking this man as a guide, he led him over mountains impracticable for a wagon road to the south fork of the St. Joseph's. Here the country became more favorable, but the Indian here meeting his brother and others of his tribe, became reluctant to lead him further, and indeed went so far as to discuss and plan mischief to Mr. Sohon, but who, fortunately understanding somewhat of their language, discovered the plot laid for his life, and with a creditable policy extricated himself from their plans. Having learned the imminent danger to which he was exposed, with two men I hastened an express to him to rejoin me, which he did at the mouth of the Mocahlissiah. His trip developed one fact, that a route in that direction is impracticable and out of the question, and I have since learned was only spoken of by the Indians in order that we should follow it, and thus expend the season in doing nothing. The Cœur d'Alène Indians, though whipped last year by Colonel Wright, are still treacherous and faithless; it is only by the most watchful and rigid course that we are not led into difficulty with them. I have enjoined upon the men from time to time the most rigid prudence, and nothing shall be left undone that will go to promote the most friendly relations with this and other neighboring tribes. The Indians have all been restless since we have been in the country, but are now becoming more tranquil. I avail myself of every occasion to impress them with the advantages that must accrue to them on the opening of our road, and, indeed, many of them begin to see and appreciate them, while others say that it is "an immense *wedge* that we are planting in their midst, that must rend them and their country asunder."

The relations with the Indians have given me much anxiety, and be assured I have left and shall leave nothing undone that shall create any but the most friendly relations with them. But under the most favorable circumstances there are always to be found some *mauvais sujets* among them. This has been evinced in the recent acts of depredations of the Pelouse, which the enclosed communication to General Harney, headquarters, and marked No. 1, will give you full details. The enclosed paper, marked No. 2, will also show you how guarded we have to be regarding the Nez Percés and their movements. Thus far, however, all things have moved on as well as we could have anticipated ; and should no untoward accident befall us, we shall yet successfully accomplished all for which we started.

Hoping to reach the Cœur d'Alène Mission by the 8th of August, we shall have the best part of three months, August, September, and October, for work in the Cœur d'Alène and Bitter Root mountains. This should, unless our work should prove too severe, bring us to the mouth of the St Regis Borgia river, and possibly to the Bitter Root river, some eighty miles from the main Bitter Root valley. In case I find that I cannot reach the valley with my train, I regard it much better that I should return to winter at Walla-Walla, sending a small party to the Bitter Root. By the middle of September I shall, however, be the better prepared to arrange my programme for the winter. Under any circumstances, captain, our appropriation should be renewed, and the remainder of the appropriation asked for by the honorable Secretary of War should be asked for in his annual estimate; and I therefore urge it upon the department, at this early day, in order that as much time as possible should be given to the subject in the early days of the next Congress.

My party thus far have been in good health.

Trusting that my course thus far has met with the approval of the department, and requesting that all communications be addressed to me at Fort Walla-Walla,

I am, sir, with respect, your obedient servant,

JOHN MULLAN,

1st Lieut. 2d Artillery, Com'g Military Road, &c.

Captain A. A. Humphreys,

U. S. Top Engineers, Com'g Office Exp. Surveys.

III.

CAMP AT CŒUR D'ALÈNE MISSION,
August 16, 1859.

SIR: My last communication to the department was from my camp on the St. Joseph's river, July 31. At that time we were engaged in building flatboats for the St. Joseph's and Cœur d'Alène rivers. On the morning of the 1st of August a flat, 42 by 12 feet, was finished, when the escort and supply train of Lieutenant White crossed to the right bank, and on the morning of the 2d of August I completed the crossing with my whole party. The boat for the Cœur d'Alène river was completed on the evening of the fourth and rowed by the men, *via* the lake, to the Cœur d'Alène river, and established at the crossing of that stream. The valley of the St. Joseph's, being somewhat wet and marshy, compelled us to build a corduroy for 400 yards across it, which, too, was completed by the evening of the 4th of August. The route thence to the Cœur d'Alène river, a distance of eleven miles, which had been explored and worked with the parties referred to in my last communication, was also completed on the morning of the 5th of August, affording us an excellent road. This had occupied fifty men for seven days in rock, earth, and timber, but we left behind us a road that we passed over with single teams with facility. Our advance working parties therefore, on the evening of the 5th of August, with the main body and escort, encamped on the left bank of the Cœur d'Alène river. Having some days previous sent a small exploring party in advance, I found that my best location would lay along the left bank of the Cœur d' Alène for nine miles and then cross. This was rendered necessary by the high rocky spurs coming to the water's edge; rock, mostly a hard limestone and slate. Selecting there a good grazing camp, our parties were again set to work, and completed the road to the crossing of the Cœur d'Alène by the evening of the 10th of August. This work consisted in labor through open timber, one mile of grading, and building one large and three small bridges over small creeks emptying into the Cœur d'Alène. The road being completed, we effected the crossing of the Cœur d'Alène in safety by the night of the 10th of August, and encamped on a small prairie on its right bank. The road from this point to the Mission being explored by the time that the work in the rear was finished, I found that we would be here delayed for some days, as the work in timber and along the scarps of the spurs or hills was somewhat severe. This road was commenced on the morning of the 11th and completed to-day, which I regard as the finest piece of road yet constructed. Here I had my whole working force, laborers and teamsters and a detachment of 20 men from the escort. The road thence being open through to the Mission, our command moved from our camp at the crossing, or a point two miles to the east of the crossing, and reached here at noon to-day. Our total distance from old Fort Walla-Walla landing to this point is 228 miles, and from the new fort 198½ miles.

In my last communication I had hoped and expected to reach the Mission by the 8th, but often we would begin the work along side hills and spurs, which on the surface appeared simply earth work, but soon we entered bed rock, which rendered our work of blasting and picking slow and tedious; but under all the circumstances of the difficulties of the route, I regard that we have reached this point of our line in good season, and possible for this latitude in midsummer. I am still sanguine as to our reaching the Bitter Root, unless some untoward circumstance, not to be foreseen, should retard our progress. On our line from the crossing of the Cœur d'Alène to this point we constructed three substantial bridges over several creeks that make from the spurs to the north of the river. Since last writing you Mr. Kolecki and Mr. Weisner, with the Mission barge, started from St. Joseph's river with their instruments, and established an observatory at the Cœur d'Alène Mission, where they have been collecting and are still collecting data for the determination of its position. We have with great accuracy the moon culminations, beginning on the 4th instant, and sets at meridian and night for latitude. Thus far two of my chronometers work excellently, the remainder are somewhat indifferent. Our barometrical observations, which have been made from the Dalles to this point almost hourly, are carried on at the observatory with great care. We have been enabled to bring our barometers thus far without accident or injury.

I find our latitude here differs but a few seconds from what we found it by our observation of last year, and the approximate longitude, as thus far determined, will throw the Mission a little to the west of that heretofore given it. Our observations, however, are, as yet, worked only *approximately*. Since last writing you, the party of Mr. Howard, lately in charge of Mr. De Lacey, has joined me and been put to work on the road. They completed their survey July 31, running their line of levels to the mouth of the Pelouse for 191 miles, the longest lines of levels yet run on the coast. Their route, as before stated, was from the Dalles, up the Columbia, on its left bank, *via* old Fort Walla-Walla, to the mouth of the Snake river; thence up the Snake, crossing it at its mouth, along the right bank, to the mouth of the Pelouse. Owing to the severe character of the work along the Snake river, heavy and precipitous basaltic rock, Captain De Lacey found the more practicable line to cross the river and keep along its right bank. The general grade along his line of the Columbia will be thirty feet, and more severe along the Snake, reaching at the maximum, at some points, of sixty feet per mile. This spirit level line has been run with much care, and reflects credit upon the party of surveyors, and will yet furnish the material to be worked up at the proper time, and turned to a practicable account. Already is it contemplated to construct a railway over the line that we surveyed from the Dalles to the mouth of the Des Chutes, over which the whole trade into the interior of Oregon and Washington will be transported.

Mr. Howard, who, at the date of my last communication, was an invalid at Fort Walla-Walla, joined me on the 14th instant, very weak and feeble, and is now quite unwell and under treatment. He is affected with a swelling in his groin, which now completely unfits him for the field. With two other exceptions of my workmen, and one or two soldiers, the command is in good health. The Indians, thus far, are friendly, and we continue to exercise, by night and by day, the strictest vigilance in our movements among and our intercourse with them. From this date, until our road shall be completed across the divide of the Cœur d'Alène mountains, my plan of operations shall be as follows: A party of twenty-three men will go to-day, with a pack train, eleven miles distant to a point of timber, and work towards the Mission; a party of forty men, with three wagons, will start this evening, cross the Cœur d'Alène at the Mission, where it is fordable, and push the work forward to join the party in advance, and I trust, by the 21st instant, to reach a prairie eleven miles east of the Mission. As additional supplies for our escort will start from Walla-Walla, and be placed in depot at this point, to be here by the 15th proximo, I have deemed the following the most judicious plan of work:

To go in advance myself from this point, with all the working parties, and push the road vigorously forward towards the waters of the Bitter Root, striking the headwaters of the St. Francis Borgia; to leave behind here in depot, until the road to the divide shall be completed, Lieutenant White, with his escort, in command of all the wagons and depot. He thus forms an entrepôt between us and the lower country, and, in command of the Mission valley, he holds and guards the key to the mountain pass, as there is nothing ahead to fear. Besides, here our large band of stock and work animals can find rich and abundant grazing; and when the road iscompleted beyond the Cœur d'Alène divide, a distance of about fifty miles, they can rejoin us in five days.

This, then, will be our plan; and when you next hear from me I hope to be well on in the heart of the mountains. The weather, during the day, is pleasant and well suited for work; the nights are cold, the thermometer falling as low as 32°, with ice $\frac{1}{4}$-inch thick.

Thus far our camp grounds have been good and grazing excellent, our animals, except those severely worked, being in good condition. I use the oxen in the woods, removing logs and timber, but generally they are in good condition. I shall send Mr. Sohon in advance to-morrow, to go as far as the Bitter Root valley, exploring and mapping in great detail all the points of the line, so that we shall know the points of greatest difficulty, and thus be enabled to arrange to the best advantage our smaller working parties. I have a detachment of twenty-one men every day, from the escort, on fatigue service—this party under the charge of the civil engineer.

The Indians and others, in times past, have told me much regarding a practicable and easy route that would lead to the Clark's Fork, at Thompson's prairie. In order to have this route examined and mapped, and thus add to our topographical knowledge of the mountain section, I shall send Mr. Engle to-morrow, with some friendly Indians as guides, over the route, and crossing the Clark's Fork river, go as far as the Flathead lake, and, returning *via* the mouth of the Bitter Root river, by the St. Francis de Borgia valley, join me on the line of work. Mr. Kollecki and Mr. Weisner will still remain, until the work is completed ahead, at their observatory at the Mission, with Lieutenant White's escort. I will also send, as soon as the water falls sufficiently low, either Mr. Howard or Mr. De Lacey, *via* the river and lake, to the upper falls on the Spokan river, to examine in detail and map the same, and, provided with the necessary tools, see if the barrier of rocks that now constitute the obstruction to the free passage of the waters of the Cœur d'Alène lake, and which I referred to in my topographical memoir of last year, cannot be removed. Could this be practicable, it would prevent the overflow of the St. Joseph's and Cœur d'Alène rivers, keep their valleys dry, and be the means of reclaiming thousands of acres of the richest agricultural land along the western water-shed of the Cœur d'Alène mountains, and will be worth all the trouble that the labor will cost in making the examination.

Mr. De Lacey will, I think, however, accompany Mr. Sohon, in exploring in advance, so that this last work will probably devolve upon Mr. Howard on his recovery from his present ill health. You were informed in my last letter that Mr. Engle and Mr. Sohon were absent on an examination up the forks of the St. Joseph's river. They examined both forks, some twenty miles above where I made the crossing of the river, and found the valley of each fork to this distance to retain the general characteristics of the main valley, where the high pine-clad spurs closed in on both sides, rendering the valley an almost mountain defile, the width of the stream itself. This examination shows the St. Joseph's to be the larger and longer of the feeders of the Cœur d'Alène lake, having its sources near the headwaters of the Pelouse, far in the eastern spurs of the Bitter Root mountains. This examination of this river has thus supplied us a topographical desideratum much desired.

You have thus an exhibit of what has been done since 31st July, and our general plan of work and movements until my next communication, which will be about 1st proximo, when I hope to be near the headwaters of the Cœur d'Alène river. Since last writing you, the Indians have given no additional signs of malevolence, or a feeling that would materially interfere with or retard our progress. An express from the settlements yesterday brought me the communication of the department, stating that $30,000 would be placed to my credit in New York, and $20,000 in San Francisco. My accounts for quarter ending 30th June, supposing the whole appropriation to be subject to my draft, would leave some seventy-one thousand dollars on deposit. My monthly expenses are now a little over $5,000 per month, and I owe Mr. Chouteau for 24,000 rations sent from St. Louis, 27th May, to Fort Benton, and freight on same, amounting in all to some $10,000, which I will pay as soon as the deposits are made. I owe also to quartermaster's department some $2,000, which covers all the indebtedness of the expedition up to date, except the monthly pay of men. The monthly account for July will go by this express, and show you the state of affairs up to August 1. The remainder of the appropriation, therefore, I would respectfully ask to be placed to my credit with the assistant treasurer in New York, as my men will be discharged on the Missouri, and it will be more convenient to have the money in New York than in San Francisco.

Although my monthly expenses are greater than originally, roughly estimated, they are not a dollar too much for our work, though it may be for the appropriation, but I have reasoned thus: should we reach the Bitter Root in good season, we shall reach Fort Benton in early summer, where I can disband a large part of my force and send them, *via* the Missouri, to St. Louis, and thus what is expended now, saved then. Besides, I am in hopes that the remainder of our appropriation will be asked for, and I can only entertain the hope that the department, understanding in detail, and appreciating the difficulties and severity of our work, and the cost of transporting provisions and material of construction over so long and distant line in a perfect wilderness, will at this early day take the subject under its special notice, and submit to the approaching Congress the estimate made by the honorable Secretary of War last year, and which was indorsed by the Military Committee in the Senate. Besides, I shall suggest, and if the views I now entertain are carried out as I could well wish, I will then recommend to the department that when I shall have completed the road to the Missouri, which I hope will be in midsummer of 1860, that I should retrace my steps over the route to the Columbia, repairing and otherwise improving the route as circumstances in the interval shall determine. On my reaching the Bitter Root valley I shall then lay before the department certain views and plans of operation that the results of our expedition have developed.

Thus far all moves well with us, and I can only hope that the same good fortune and success that has attended our movements to this point will continue to the end of our expedition.

Trusting that my course and plan of operation thus far have met with the approval of the department,

I am, captain, with respect, your obedient servant,

JOHN MULLAN,

1*st Lieutenant* 2*d Artillery, in charge of road.*

Captain A. A. HUMPHREYS,

United States Topographical Engineers,

In charge of the Office of Explorations and Surveys.

IV.

CAMP ON CŒUR D'ALÈNE RIVER,

21 *miles east of Mission, September* 4, 1859.

SIR: My last communication to you, dated at the Cœur d'Alène Mission, 16th August, gave you up to that date our progress and general movements.

In accordance with the programme then given, I moved all the working parties forward in four sections, working the road from the Mission up the valley of the Cœur d'Alène towards the summit of the mountains, or divide separating the waters of the Cœur d'Alène from those of the Bitter Root, leaving behind me my wagon train with the escort under Lieutenant White, to guard our rear and keep open the communication between Fort Walla-Walla and the Mission, and at the same time to give the necessary protection to our train.

Each of these parties moves with a pack train, which is necessary in order to take advantage of the sparse grazing, and at the same time to expedite our general movements. As soon as the route was opened from the Mission to some eight miles eastward, we struck an open wooded country, which afforded sufficient grazing for two weeks for our large band of animals. Lieutenant White, in general charge, therefore, established here a depot camp. The grazing having become nearly exhausted, he will move to-morrow to a point ten miles east of the Mission, where he will find sufficient grass for our train until I can open the road to the foot of the mountains, where again we strike a large prairie that will afford us grazing, say, for ten days or two weeks.

Having, therefore, in my judgment, deemed the plan proposed the best under the circumstances, the parties have moved forward, working the road through the timber, and last night (Saturday, 3d) reached with the rear party our camp, twenty-one miles east of the Mission. There are two parties still in advance, and by the 10th of September I hope to reach the prairie near the foot of the western water-shed of the mountains, and by the 20th September to reach the summit of the mountains, and probably by the 15th October to strike the Bitter Root river.

The work thus far has been heavy, and hence tedious. The timber is large and dense, but I can report that we have left behind us as excellent a route as any mountain region can boast of. The work has consisted of clearing of timber, grading and bridging. It would be almost difficult to say how much bridging is required, but suffice it to say that every point has been bridged or corduroyed that required it. My rear party moves with four wagons and our forge, and they have thus far passed over the route without difficulty or accident.

The general direction of this valley is as represented by the last maps of Governor Stevens, but the spurs make down upon the river, overlapping each other, rendering the stream itself very tortuous, and hence to be frequently crossed. All the crossings are now fordable, and are always fordable from 1st June. Owing to the number and character of these, I have deemed it, with my present time and appropriation, a matter of complete impossibility to stop to bridge them now. Already up to this point we have made fourteen fords, each involving a bridge of heavy structure, that would require not less than two weeks to each, or seven months' bridging alone. For all practical purposes, as this route will and can only be travelled when the grass is in season, our present system is the best. From the Bitter Root river to Fort Benton, as we shall have, say, December, January, February, and March, I shall be enabled to get out bridge materials for each of the crossings of the streams, so that it can be travelled in high as well as low stages of water. But at present it is essential that we open the road through the heavy timber, cross the mountains, and reach a wintering cantonment in the Bitter Root valley or its vicinity. What I would propose would then be that, pushing on with my party to Fort Benton, say, in early summer, I will there await the arrival of the steamer from St. Louis; and in case we have our appropriation renewed, I can have sent me from St. Louis another party of workmen, hired at reduced wages, and, returning over the route, render it of a more permanent character. To build a line of six hundred miles of road, at an average of $200 per mile, in a country where the lowest price of labor is $50 per month, and transportation at the rate of from $200 to $300 per ton, is less than roads have ever yet been constructed; and the difficulties and length of the route, and the manner of overcoming them, have been so often and in such detail set before the department in my previous reports that it would hardly seem necessary to call attention to them again.

But in view of the great length of time required for my communication to reach Washington and a reply to be had, I would again earnestly recommend to the department that our appropriation be continued, say, $100,000, in order that we may creditably complete a work now in successful progress, and the results of which are to redound at an early day to the best interests of the country. My monthly expenses are now near $6,000. I owe to Messrs. Chouteau & Co. over $10,000 for stores and transportation to Fort Benton, and my employés, up to 31st August, over $15,000, making, say, $25,000, and balance on hand August 31, say, $68,000 of appropriation, leaving, say, $43,000, and at the rate of $6,000 per month in seven months, or from 1st September to 31st March, or in April, our appropriation nearly exhausted, at which time I shall have to discharge all my employés at Fort Benton, and send them in boats, *via* the Missouri, to St. Louis or the settlements. In view of these things, therefore, I trust the appropriation will not only be asked for, but at such an early day that we may continue our work in the field.

An express reached me on the 18th of August from Fort Benton, giving me the gratifying intelligence of the safe arrival at that point of the staunch little steamer Chippewa, having on board some 24,000 rations for my party, and such articles for my road as I needed; all safely delivered by steamer direct from St. Louis to Fort Benton. This, therefore, is a complete success, and practically proves that which we have so often advocated. With the success in steam navigation had on the waters of the Columbia during the present year, (for steamer Colonel Wright has navigated up to mouth of the Ton cañon at the point where our road crosses the Snake river,) and with steamers to Fort Benton, it needs no special proof to show what bearing our present labors are having in connecting these two heads of navigation on the two great arteries of the country. Already is the result of our labors being appreciated, for travel is already had over our line, and much more contemplated. When opened, the route can be made by troops with easy marches in sixty days; and as soon as we shall have reached the Bitter Root, I will be enabled then to lay before the department, with confidence in their final success, certain movements that can be carried out, demanded by the military necessity of the service, and that can be effected with a saving of 50 per cent. to the Treasury Department of the amount that it would cost by any other route.

Mr. Chouteau will, in all probability, put another steamer on the river early next season, and I hope to reach Fort Benton at an early day to meet any party on her arrival.

Since I last wrote you, I sent Mr. Kollecki to examine the country from the Mission to Fort Colville, where a large post has been recently established, with which a wagon road is had to Walla-Walla. This

I deemed of importance, not only in a topographical light, but with a view of collecting data and information that would enable us to judge of the practicable connexion by a wagon route for military purposes to that point. I will report the results of his exploration in my next letter. I am confident, however, that a good wagon road line can be had, and when the military necessity arises for connecting this point by our road with points east, we shall be possessed of all the information requisite.

Mr. Sohon and Mr. Engle are still absent, the former selecting the best line of location for our road hence to the Bitter Root, and the latter exploring and mapping a new route recently discovered by the Indians from the Cœur d'Alène Mission to the Clark's Fork, at Thompson's prairie. I had an express from Mr. Sohon from the summit of the Cœur d'Alène mountains, up to which point he had furnished me with sketches of the route on a large scale for our working parties of the whole route from the Mission to the summit.

Both Mr. Engle and Mr. Sohon have with them a party of friendly Indians as guides and employés. From the results of their labors I expect much. We shall lose no opportunity to collect all available data that will go to form a complete and correct map of this region. As I informed you in my last communication, my observatory was established at the Cœur d'Alène Mission by the 3d of August, at which point we have had daily and continuous observations made up to date. To-day I shall send instructions back to Mr. Weisner and Mr. Kollecki who have had charge of the observatory, and direct them to discontinue their observations at that point and establish their observatory at the foot of the Cœur d'Alène mountains, with a view of fixing accurately the position of the pass across the Cœur d'Alène mountains. This point well determined, and the juncture of the St. Francis Borgia with the Bitter Root, will be excellent intermediate points between the Mission and the Bitter Root valley.

Be assured that we shall leave nothing undone that shall be necessary to mark and determine every point that will be necessary to refer to at any time in the future. The region of country through which we have passed from the Mission eastward is a dense forest of fir, pine, hemlock, large cedar, spruce, and alder, and near the margin of the river cottonwood. The valley has a general width of from $\frac{1}{4}$ to one mile, with spurs, however, often coming down close to the water's edge, and within 300 yards of each other. The general grade of the valley is, say, 25 feet to the mile. In most places the geological formation is of limestone; in many places, however, we have struck large veins of quartz between huge beds of slate. The slate formation is after the limestone more frequently met with, and from time to time large beds of granite. The indications of gold are frequent, though only in one case have we found scale gold. The color and black sand is frequently met with, and I avail myself of every opportunity to have the country prospected and explored, as I have in the expedition many old California miners.

The Indians thus far are quiet, and I shall endeavor by a just and prudent course to retain their friendship. They have quietly settled down regarding our road, and now, instead of regarding it as a cause for dissatisfaction, rather look upon it as an especial advantage, and make use of it instead of their Indian trails.

We allow no trading whatsoever with the Indians, unless it be done with an interpreter, and have, in a word, taken every precaution to avoid either a difficulty or a conflict with them. They are now at the many points in the country collecting roots, fish, and berries, and cease to be alarmed at our continued presence in their country.

The Jesuit fathers *now*, I think, use their influence in behalf of our movements; but truly, I am led to believe that much of the hostility and opposition heretofore had on the part of the Indians against our road has originated in the feeling of lukewarmness, if not opposition, direct on the part of the fathers to our work and to our road. I am loth to come to this judgment and conclusion, and it is with regret and reluctance that I make it, for personally, the fathers have been kind to me and attentive to my every want. But there are evidences so strong that in my official capacity I have been led to observe them, and hence report them; and though it may seem strange, yet I confidently believe that the conflict with our troops that led to Steptoe's disaster can be traced to causes that had their commencement in that feeling which has taught all missionaries to keep the Indians aloof from contact with the white man. The charge of inciting the Indians to an open conflict would be serious in itself against any white man, or band of white men, but still there are evidences enough to lead me to suppose that had the missionaries not been in the country, we should not have been compelled to record a reverse to our arms, and a disaster that in itself has rendered the Indians bold, saucy, and impudent. Since I have been in the country I have told them all what their punishment shall be in case they destroy any of our boats, bridges, or improvements, and which truly I shall carry out if the occasion arises, which I am happy to say that thus far it has not: namely, that I shall hang every man who, forgetful of his promises made to Colonel Wright last summer on the conclusion of the war, shall dare to depredate upon our road. I hope to write the department again between this and 20th instant. I received per last express a letter from the department informing me that a deposit of $30,000 would be placed to my credit in New York, and $20,000 in San Francisco. By the summary statement rendered August 31, which, with other papers, is herewith enclosed, a balance of some $18,000 is due the United States, which, with $50,000 yet to be placed from the appropriation, will be $68,000. My liabilities up to date are $25,000, leaving $43,000. I would, therefore, request that the remainder of the appropriation, $50,000, be placed to my credit with assistant treasurer in New York city, to meet contingencies as they arise.

Trusting that my plans of operation and general movements thus far have been approved by the department, I am, sir, with respect, your obedient servant,

JOHN MULLAN, 1*st Lieut.* 2*d Art.*,
Charge of Military Road from Fort Walla-Walla to Fort Benton.

Captain A. A. HUMPHREYS,
U. S. Top. Engs., charge of Office of Exp. and Surveys.

V.

MILITARY ROAD EXPEDITION,
Camp on north fork of Cœur d'Alène river,
30 *miles east of Mission, Sept.* 19, 1859.

SIR: My communication of 4th September, from camp 20 miles east of Mission, informed you of my progress and general movements up to date; since then, though our progress has been slow, our work has been great. From that date to the present, the work has been in large and heavy timber—cedar, hemlock, white pine, and fir principally; and in places, much and heavy grading in order to secure the best location. The timber has been heavy to cut, and equally so to clear. On the 12th instant we reached the forks of the river and the forks of the road, and finding the best location across the mountains by the northern fork, our work since then has been along it. We made sixteen fords or crossings on the main stream, and up to date have made eleven on the north fork—in all twenty-seven crossings. You will form probably a better idea of the character of the river and the formation of the mountain spurs from this than any other description I might be enabled to give you. As we enter mountains, as we anticipated, the work becomes heavier, and the spurs of the mountains overlapping each other turns or drives the river from bank to bank, rendering it exceedingly tortuous. There is no alternative left but to cross the river, or have heavy rock cutting along points of spurs; the latter consumes more means than is at our disposal, and the former hence has to be done—the river forded and not bridged—for the cost of time for twenty-seven heavy bridges would run us far into the winter. Therefore you will readily see that the views contained in my last report are the only ones that can be practically carried out. From this to the mountains I shall have not less than fourteen more crossings; on the St. Francis Borgia not less than forty-six; each of these, in order to render the route in a travelling condition in the stages of high water, involves a bridge—some small, many large—and hence it needs no reference on my part to set before the department the time and cost requisite to construct them. It will be well if, with our present appropriation, we can open the route to the Missouri through the timber and mountains, and render it practicable for either an emigration or military purpose at seasons of low water. But if these bridges are built, it can be travelled at all seasons, and hence my original plan and the plan indorsed long since by Governor Stevens, who is equally conversant with the character and the difficulties as myself. My plan, then, is, and which again I suggest and with respect submit to the department for its consideration and a claim thereon, and hence recommendation to Congress, that the route be opened to the Missouri, where, being already provided with the tools, material, and means of transportation, which we have now abundantly on hand, and with a party of workmen hired at reduced wages and brought from St. Louis, we can retrace the road and put it in a complete and practicable condition. I trust, therefore, the department, appreciating what has been done, and the importance of the road in a military point of view at the present time, when our difficulties on the North Pacific are of a character that might well arrest our attention, will, by its action and favorable indorsement, bring to a successful end a work now in successful progress.

Since last writing you both Mr. Engle and Mr. Sohon have returned from their successful explorations. The former reports in detail the character of the line, and represents everything as favorable on the proposed line of location as we could desire, and no difficulties so great but what I trust our zeal and industry may be found equal to cope with. Mr. Engle's trip has definitely settled the character of the new route hence to the Clark's Fork. He passed over the divide separating the waters of the Spokane from those of the Clark's Fork, and found the route difficult and uninviting and impracticable, either for a wagon or rail route. His exploration developed the country at the head of the north branch of the north fork of the Cœur d'Alène river, and has added materially to our topographical knowledge of the region between the two streams above referred to; returned *via* mouth of BitterRoot and St. Francis de Borgia and across divide along the southern fork of Cœur d'Alène, and secured in detail certain topographical material much needed in that section. Mr. Sohon examined in his trip both banks of the Bitter Root; found we would cross it three times in order to secure the best location and avoid difficult work; unfortunately they both represent the whole country burned and the grass destroyed. I am somewhat anxious regarding the welfare of our stock, but trust the September rain will cause the grass to spring up. I regret to state too, that the weather has recently been unfavorable for our progress—several rainy days in succession; but to day the clouds have broken, and we trust the bright sun may be the harbinger of future good weather that will materially add to our progress. Mr. Kollecki has also returned from his trip to the Colville wagon road, and made in great detail maps and cross sections of the falls at outlet of Cœur d'Alène lake, made with the view to ascertain if a removal of the fall would not materially change the volume of the water in the river and lake. This was referred to in my topographical memoir submitted to the department in February last, which please see. He confirms my views as then given, and finds that the overflow of the fertile valleys of the Cœur d'Alène and St. Joseph can be prevented. This I regard an important consideration in the future settlement of this country, and will be the means of reclaiming thousands of acres of most excellent agricultural land. I had intended to apply the services of Mr. Howard to this special work, but I regret to report his continued sickness, and his inability to perform any service whatever since the middle of June last; and I find my force so completely occupied that I have not a single man to spare for the work, but the material collected will at the proper time be arranged and laid before the department for the action of those who may yet retrace our steps, provided it be not yet accomplished by ourselves; for we are anxious and willing to do each and everything in connexion with our movements that need add to the development and the knowledge and information of the region in which we are laboring. Mr. Kollecki is now in charge of a party en route, mapping in great detail the line of our route from the Mission to the mountains. Mr. Weisner is with this party observing for time and latitude, and at the same time running a barometrical line to compare with a line of spirit levels that is being run by Mr. Johnson. As the principal difficulties of the whole route from Fort Walla-Walla to Fort Benton have always been regarded by Governor Stevens and myself as lying between Cœur d'Alène Mission and the Bitter Root river, I am desirous of laying before the department in detail and with accuracy, an excellent map and a complete line of barometrical and spirit levels through the pass over this route. This line I shall continue to Fort Benton. Our level line from Fort Dalles to the mouth of Pelouse, one hundred and ninety-two miles, was run very satisfactorily, and I only regret that the number of my barometers was such that I could not spare one to run in connexion with the line, and then ascertain the exact reliance to be placed on our barometers, and at the same time lend our aid in adding more data to barometrical science. We shall however avail ourselves of the opportunity that the present occasion affords us, and collect all data that can be made of availability hereafter. I find in Mr. Weisner a gentleman of marked talent and ability. His labors in the astronomical department are untiring and satisfactory. In addition to securing material for the latitude and longitude of each point requisite, he finds ample time to devote to many points that add to the knowledge of the scientific world. While our observatory was established at the Mission he more particularly devoted himself to observations on the spots of the sun, the results of which will, at the proper time, be laid before the department. Regarding Mr. Kollecki's trip to the Colville wagon road, I would state he found the route practicable, but at heavy cost from the Mission to the Cœur d'Alène lake, provided the north bank of the Spokane be followed. But from my knowledge of the country and the character of the Spokane, as shown by his trip, I am led to suppose that the south bank for a certain distance will be found more practicable, and by crossing at a point known as "Antoine Plants," that a good line can be secured. Mr. Engle is now at leisure for this trip, and to-morrow I shall send him forward to its examination and trust to lay the results of his trip before the department in my next communication, which will be about the fifth of October. From our slow progress during the two weeks past we shall fall far short of the times of arrival at certain points of the route in advance, referred to in my last communication. But without referring to any special dates, for there are so many contingencies daily arising that render our calculation very uncertain, the department may be assured that our road shall be pushed forward with all the energy, zeal, and industry that our party is possessed of. I would here simply state that though yesterday was Sunday, our men were diligently at work on the road in order that we might avail ourselves of every pleasant day before crossing to the summit of Cœur d'Alène mountains. A supply train for the escort reached Cœur d'Alène Mission without accident or difficulty on the 15th instant under Major Grier, making the trip from Walla-Walla in *thirteen days*. The department will, I trust, appreciate our labors on that part of the road alone when they remember it cost our party fifty-one days to open it.

The Indians still remain quiet; the health of our party is generally good; two men, Rease and Dickinson, have been discharged on account of ill health; some accidents have befallen our men, that I trust time and care will remedy; two men cut with axes, one man accidently shot in the knee, and one injured by a falling tree; Mr. Howard still in bed from swelled groin. The remainder of the men are at work each day. Four men, Barstow, Lyne, Brennan and Irons, were discharged on the 16th instant, as they did not come up to the standard and requisites for laboring men on this expedition. I sent them back to Walla-Walla with Major Grier's train, provisioning them up to the first of October. These are the only men I intend discharging for the present, as my force is the smallest possible to accomplish the work intrusted to me. My route hence from this point will be up the valley of the north fork following generally the river to an open prairie, five miles from the summit of the mountains, which is now eight miles distant, and if we reach the eastern base of the mountains by the 5th or 10th of October, I shall rest satisfied, for then we shall be secure from snow in the mountains. I shall write the department again by the 5th of October, at which time I will forward returns for the month of September and for the third quarter. I have this day forward to P. Chouteau & Co. a draft for ten thousand six hundred and eighty-five dollars and twenty-five cents ($10,685,25) for stores and transportation to Fort Benton, and have directed them to forward one copy of the voucher to you and the other to myself for the reason that they will be included in the third quarter, and I send them the vouchers by this mail for their signature. Since the time occupied for a return will be too great, and as winter is approaching, we may be much delayed in receiving our communications, &c. The vouchers will be retained till my accounts for the third quarter are received at your office, when they will be included in the same, as will be hereafter referred to.

Trusting that my course and plans of operation will meet with the approval of the department,

JOHN MULLAN,
1st Lieutenant 2d Artillery, in charge of Military Road.

Captain A. A. Humphreys,
U. S. Top. Engineers, in charge of Office of Exp. and Surveys.

VI.

Military Road Expedition to Fort Benton,
Camp at foot of divide of Cœur d'Alène Mountains,
October 4, 1859.

Sir: My last communication, dated September 19, gave you a general journal of our movements and purposes, since which date we have continued through the same character of country and work up to date. We reached our camp at the foot of the divide on the night of the 28th September, and all our force is now working vigorously up the western slope of the divide of the mountains. After a long and careful examination of the range and its different gaps and depressions, we have found a pass, which, probably, is the lowest in the Cœur d'Alène range, and which, in honor of Mr. Sohon, who made the first topographical map of it in our expedition, I have termed "Sohon's Pass." It is a low (comparatively) saddle between the high mountain peaks, and within one-half a mile of what is called by Governor Stevens "Cœur d'Alène Pass." The eastern slope of this divide is most beautiful for our purposes; does not involve *one foot* of grading, is slightly timbered, and, probably, our wagons may descend it with wheels unlocked. Our road over it is entirely on the western slope. It will be by our road about one and a half miles from base to summit, and as the slope is too steep, will involve two curves along the spur, which will cost heavy work in grading; and as all the timber on the slope is heavy and hard, the tedious and more difficult work will be the grubbing out the large stumps, the roots of which, at times, would seem to run over the whole mountain side. This will occupy us till the 12th instant, and I hope by the 15th instant that our advanced working parties will be encamped on the St. Francis Borgia. From that point to the Bitter Root, about thirty miles, I shall push onward with all the speed and vigor possible; for I am well aware that the season is far advanced, and the weather, in the shape of ever incessant rains for two weeks, has been much against us, and though it did not suspend the work, it has retarded its general progress. As soon as the work over the divide is completed, I shall begin to move my train of wagons over, now some twenty-eight miles to our west, and in consequence of scarcity of grass in the mountains, shall be compelled to drive my stock to the Bitter Root river. Sparse grazing may be found west of the divide, but I fear keeping my stock west of the range too long, for the weather already is becoming very cold, and our stock, principally mules and horses, have had much service for the last five months, and many of my oxen are reduced in flesh. All these circumstances, therefore, prompt me to keep them east of the range, and as soon as the road to the Bitter Root river shall be opened, my wagon train will follow. This, then, as you will see, will bring us into the main Bitter Root valley not before the middle of December, at which time our stock will need much rest and good grazing; and I can only entertain the hope that the winter may be both late and mild. I have, since the commencement of the work, pushed forward with all the energy and will at our disposal; and you will, I trust, readily appreciate best the character and amount of our labor from the time and force employed upon it. Every teamster, cook, laborer—in fact, my every man is daily, with pick, shovel, and axe, on the road, and thirty-three enlisted men on fatigue duty in addition. You will also see how impossible it was to give the secretary, *in advance of our detailed work*, any statement regarding programme and force employed, as the principal difficulties lay between Walla-Walla and Bitter Root valley; and as the latter point is the *only one* where we can safely winter, you will, I trust, appreciate my plan by pushing on this section of the work with the largest force possible. But this will involve our appropriation being totally expended in spring, or very early summer, and I can only entertain the hope that the department will take those steps as will continue our appropriation. We are now 267 miles from our initial point, Fort Walla-Walla, and as we progress the length and difficulty of our communications are increased. Up to the present time my mail through the Indian country has been carried twice a month, (there being no mail route,) by two of our men. From 1st November I shall make the attempt to establish winter express once a month. For this purpose, I propose establishing small depots of oats for forage at Walla-Walla, at Snake river, at Cœur d'Alène Mission, at divide at mouth of St. Francis de Borgia, and Bitter Root; and, with fresh horses, hope to be in communication with the department once a month. If this is not done we shall be cut off for six months from all communication, and I regard the interests of our work such that I feel warranted in pursuing the plan indicated, both at the risk of expense, difficulty, and danger. My levelling party is behind only 12 miles. They find easy grades from the Mission, commencing with 12 and ending, at the latest dates, with 25 feet to the mile. I am confident that our late results, and the detailed topography along the northern slope of the mountains along left bank of the Cœur d'Alène river, will develop such a line as to show not only easy grades in our approach to the mountains for a railroad line, but on entering the pass, will involve a tunnel of not more than one-half mile The detailed level and topographical line will, however, show it beyond cavil or doubt, and I regard the matter a discovery worthy of special note in the railroad question. Mr. Engle, whom I sent forward on the Colville line, has not yet returned. I look for him daily, and will report the results of his labors in my next communication.

I will establish an observatory on the summit of the pass on Thursday the 5th instant, where, with our transit, we will observe the moon culminations for the coming lunation, and I am sanguine to believe that with clear weather we shall be enabled to obtain good results. Our barometric line still continues to be run with our line of spirit levels. Our observations for time and latitude continue every day with the levelling party under Mr. Kollecki. This, then, will give you a general idea of our work now in progress or contemplated; and this plan will continue till reaching main Bitter Root river, where circumstances or contingencies of the moment may suggest a different one. The Indians are still quiet and molest us not. The major portion of the escort, however, is in my rear, with the wagon train, 28 miles. Our advance working parties are all still moving with small pack trains, which materially adds to the progress of our work. The wagon forge and a small depot of supplies in wagons, however, still brings up the rear day by day. The forge is in daily use, making and repairing tools and the general outfit of the train. I stated in my last report that supplies forwarded to Fort Benton would be taken up on the third quarter, but as the papers were made mostly out, they will not appear till the fourth quarter. I have already, however, forwarded to Messrs Chouteau a check for over $10,000, which will be the summary statement for October. I had intended sending a pack train to Fort Benton for supplies there in depot, but I find that I cannot spare a single man from my work, (the teamsters and packers are all laborers on the road;) and I have effected such arrangements with Lieutenant White as will not necessitate sending a train forward before reaching Bitter Root valley. I trust, too, in this the department will appreciate my plan when I shall show that the cost in transportation by the two methods or routes, if applied, will be as one to two. I shall write the department again 1st November; party is in general good health; Mr. Howard is still unable to do any duty about room; man shot is recovering; the men cut with axes and hurt by falling tree recovering; all other men daily on road.

Trusting that the department will still find our determination and energy in our work in the end, as it has in the beginning, and that my general plan will meet its indorsement, I am, sir, with esteem, your obedient servant,

JOHN MULLAN,
First Lieut. 2d Artillery, in charge of Military Road.

Captain A. A. HUMPHREYS,
U. S. Top. Engineers, in charge of Office of Exp. and Surveys.

VII.

CAMP AT FOOT OF BITTER ROOT MOUNTAINS,
October 26, 1859.

SIR: An Indian en route to Walla-Walla enables me to drop simply a line. Our road across to Bitter Root mountains is completed, after a laborious and severe work. But it speaks for itself. The road is now opened to a point five miles east of the range, and all our wagons, stores, and stock are east of the range. I shall push vigorously forward to the Bitter Root river, which I find I cannot reach before 20th November; and cannot reach Bitter Root valley before 15th December or 1st January. I truly regret that so late a season should find us hard at work with the road to the Bitter Root still incomplete; but I trust the department will appreciate the character of our labor when they learn that on the mountains alone I had one hundred and ten men at work in order to get over in time, and our work cost us twenty days at times in drenching rains. Our last train of wagons crossed the divide to-day with no snow, but the nights are cold, and I regret to report the loss of most of my horses and a number of oxen and beef cattle. The horses were Indian horses, and being much used in the summer for various purposes had very sore backs, and during the cold nights with scanty food they could not stand the cold but died by the wayside. This has grown out of the peculiar nature of our work, where they have been kept in the mountains on sparse grazing for near three months. We have had four weeks incessant rain, which rendered our road very wet and muddy, and when moving our trains from the depot camp we experienced great difficulty and delays. My depot while the road is being worked is now about five miles east of the divide, and will there continue till we reach Bitter Root river, when it will move down. When my teamsters are with my train, that much effective force is taken off the road. I have every man possible on the road, including between thirty and forty soldiers. If anything, my force is rather too small than too large for the work to be done, and, as I before informed the department, my appropriation will be expended by April or May next, and I can only hope the department will press urgently the matter before Congress, and not leave me in the mountains with my hands tied. I would ask again, in case it be not already done, that the remainder of the road appropriation be placed, as I urgently requested, with the assistant treasurer in New York to meet my drafts. I have had already one check returned protested, and will hereafter draw on New York and trust the department will there place sums to cover my drafts. The department may rely that all shall be done that shall insure our *final* success. But midwinter will, I fear, still find us west of the Bitter Root valley. I will write again from Bitter Root river.

Truly, your obedient servant,

JOHN MULLAN,

1st Lieut. 2d Artillery, in charge of road.

Captain A. A. HUMPHREYS,

U. S. Top. Eng'rs, charge Office Explorations and Surveys.

CAMP ON ST. FRANCIS BORGIA RIVER,

November 14, 1859.

SIR: I have the honor to report the arrival of my expedition with the train of wagons, &c., at a point on this river 20 miles from the Bitter Root. The road is completed to a point within ten miles of the Bitte Root. On October 20 the train crossed the Bitter Root range without snow or difficulty, and by 1st November reached a point 5½ miles east of divide; on 1st November it began snowing, and continued a perfect snow storm till 5th; with an interval of a day it resumed with great fury, and attained with our rear camp a depth of 18 inches. Our work still continued ahead, and on the 8th it turned suddenly cold, thermometer below zero, grazing covered up, and little or no browsing for the animals.

Here it beceam necessary to drive all the animals to the Bitter Root river, where they are at present, though I have lost, unfortunately, a great number, part from straying, most broken down, want of grass, and cold weather.

I await a change of weather. If it should change for the better in ten days, I shall try to reach the crossing of the Bitter Root river, but the probabilities are against me, and hence I may drive all the animals left to Bitter Root valley for winter, and try to resume work by 1st April, and reach Fort Benton by August.

I shall try to avoid a deficiency by working with a reduced force. But the department will see from the last summary statement rendered what the condition of things would be, and hence I shall push on the work to completion, relying fully upon a favorable action by the present Congress. Appreciating as I do, therefore, our own condition, and the ultimate success of our labors, I shall push forward with all the energy possible the work to a successful completion, whether it involves a deficiency or not. As soon as circumstances shall have matured my plans I shall despatch an express on snow shoes to Walla-Walla, at which time the department shall be informed in detail regarding our final condition—knowing the following facts to exist: I have an abundance of stores for my party until we can resume work; the snow now is eighteen inches deep, the river crossings frozen mostly, the ground frozen and difficult to work our animals on the Bitter Root river, and all that are left will go to the Bitter Root valley to winter. Mr. Engle has returned from Colville, and has found a good wagon road connexion to that point. He started on the 8th instant for Fort Benton with a small advance party, and will return January 1, 1860.

The department may rest assured that all things will be so arranged that the best interests of the department and the ends in view be attained. The probability is, we shall winter on the St. Francis Borgia 10 miles from the Bitter Root.

I am, truly, your obedient servant,

JOHN MULLAN,
2d Artillery U. S. Army, charge Military Road.

Captain A. A. HUMPHREYS,
Charge of Office of Explorations and Surveys.

IX.

MILITARY ROAD EXPEDITION,
Cantonment Jordan, St. Regis Borgia River, W. T.,
January 3, 1860.

SIR: My last communication to the department was dated on St. Regis Borgia river, November 14, 1859, since which time circumstances have determined our condition for the winter. When the last express left our camp the depth of the snow was such as to cover up all the grass and browsing from our animals, compelling me to drive them to the Bitter Root river. Being still on the St. Regis Borgia, and *above* some sixteen of its crossings, I feared that should I remain at the point where the snow had caught me, in the opening of spring, when work might be resumed, the freshet might be so great that the beginning of effective work would be pushed far into June. In order, therefore, not to be caught by this contingency, I set the men to work and constructed twenty hand sleds, carrying 600 pounds each, and though the river was opened to within ten miles of the Bitter Root, there was much yet to be done on it. We therefore completed it properly below *all the crossings* of the St. Regis Borgia. My animals, on the completion of the road, being still on the Bitter Root, I made the attempt to get them up and move down my train to a point below all the crossings, where I could safely winter my party. In this I succeeded for my civil party, but my animals were not able to move down the train of the escort. I lost some animals in the attempt; the remainder I started for the Bitter Root river, where I organized a small but efficient party under one of my wagon-masters, Mr. Calwell, and sent the whole band to the Bitter Root valley to winter. The ice had already formed in the Bitter Root river, and some of the animals crossed on it, and while crossing, the ice broke up, and I came near losing my entire band; with the exception of one or two being drowned, they crossed in safety, and started for the Bitter Root valley on the 9th of December. As soon as this was done, foot bridges were constructed over the sixteen crossings of the river, and the sleds were set to work to move the whole escort train also to the point that I had now selected for our wintering point, for it was now evident that there was no possibility of making the road any longer to advantage. The men had worked amid snow and cold till it was out of the question to work any longer, and on the 5th of December, with reluctance, I suspended work on the road. The work of sledding now began, and we have succeeded in moving down the whole escort train a distance of eighteen miles with hand-sleds, in all some fifty tons of freight, and are now moving down the wagons, and by 12th of January my two entire trains will be at the point where I now write you. Having selected a point on the left bank of the St. Regis Borgia, and about 10 miles from its mouth, where the mountain formation rendered the spot sheltered, and where timber and water were convenient, I set the men to work building first a storehouse for my subsistence and other property, and then erected an office, in which I am now working up all the material collected during the last eight months. The weather has now become cold, and on the 5th of December the thermometer was 42° below zero. Having completed the storehouse and office, I set the men erecting log cabins, and to our winter home I have given the name of "Cantonment Jordan." By the 1st of January my buildings were completed, when I told all the men who were thus thrown out of employment by the suspension of the work that their pay would come out of the renewed appropriation, which I fully rely upon during the present Congress. Some five in number were not willing to trust to the contingency, and preferred encountering the dangers of the snow to Walla-Walla, whence I have discharged them and paid them off. We are now some sixty miles east of the Cœur d'Alène Mission, and I am sanguine to believe that our worst work is over. I have much under-estimated both the amount and character of the work to be done on the road, and though I feel somewhat disappointed in my own person that I did not reach the Bitter Root river, still the amount of work left behind us sufficiently attests what has been done within the last six months.

I shall resume work in the earliest days of spring, by the 1st of April at furthest. I have had examinations made of the line of location to the crossing of the Bitter Root, and from the best data, I estimate that it will cost us one month's labor to reach it by the 1st of May. In November last I sent Mr. Sohon, my guide and interpreter, to examine a new line from the crossing of the Bitter Root to the Hell-

Gate, which had been described to me by Indians and others, and which would involve only one crossing of the Bitter Root. The line examined in August last involved three crossings. Mr. Sohon had with him a Flathead guide, and spent several days in its examination, but the ground was then covered with snow to the depth of fifteen inches, rendering it out of the question to pronounce definitely on the line of location, but the probabilities are that this will be the line of the road, and which at most may involve two months of work; we shall have then three months' work to the Hell-Gate defile, and I shall endeavor to complete the remainder from Hell-Gate to Fort Benton in another month, reaching the latter point by the 1st of August, provided I am not detained by high water on the Hell-Gate river; and provided further, that contingencies which I cannot now foresee shall not occur in the interval.

Every man, therefore, whose services I can possibly do without while the work is temporarily suspended, will be paid out of the renewed appropriation, so that I shall be enabled to complete the road with the remainder. I have preferred doing this to falling back upon Walla-Walla and not be able to resume my work in June or July next, whereas now work of a necessary character (office work) is going on, and the men on hand to begin work in the spring.

But I trust, captain, the earnest requests contained in my former communication have met with at least a consideration; and that an estimate has been already submitted by the department to Congress of the amount necessary to complete the road. By reference to all the roads made in this region over that line, you will find the average per mile has varied from $400 per mile, as a minimum on the Steilacoom and Monticello road, to $1,000 per mile, as a maximum on the Salem and Astoria road; and when the department will consider the length of the line over which subsistence and other stores have to be transported, and the character of the mountain work, they will not hesitate to urge with earnestness a renewal of my appropriation. But the department may rely on one thing: as I was sent out to build and construct the road, it *shall* be constructed.

The following work is now going on, all of which is necessary for the accomplishment of our special objects. The office work is being pushed ahead with all the vigor possible; and I send by this mail copies and tracing of those parts which, being partially completed, may prove of the most value for present reference—the work on the different lines examined during the past summer being made on a scale one inch to a mile, and all details put down ready for a completion of our general map. A small party is now at the crossing of the Bitter Root river whip-sawing lumber for the building of boats for the crossing of this river in the spring. At the same point I have established my observatory, which will continue during the winter; we have already two months' transit observations at the same point, and shall be enabled to determine the points of our line with accuracy. A party is in the Bitter Root valley with my wagon-master, in personal charge of my animals.

Mr. Engle, who started for Fort Benton with a small party to map the line thence in order that I might arrange for the spring work, has not yet returned. He has been absent since 8th of November. I had hoped that he would have been here before starting my express, and I can only suppose that he feared trusting himself back, and must be in the Bitter Root valley. If he should return before spring, you will be informed in my next express.

I sent an express to Walla-Walla on the 14th of November, and on 20th of same month sent an additional expressman to the Cœur d'Alène Mission to await there the arrival of the mail and to assist them across the mountain. My expressman, Toohill, tarried at the Mission until the 27th of December and learned nothing from Walla-Walla; and the fears entertained were that the expressman and a number of other men who were discharged from the escort were overtaken by a snow storm on the plains, and the men and mails both lost. I trust that this will not prove to be the case, but such is the impression, as written me by the Jesuit fathers now at the Cœur d'Alène Mission. My expressman, remaining at the Mission until the 27th of December, started for my camp with the Cœur d'Alène Indians on snow shoes. They reached my camp on the 2d of January, in five days, a distance of sixty miles. They report the snow on the divide of the Bitter Root as five feet, and most of the crossings of the Cœur d'Alène frozen over, and also many crossings of the St. Regis Borgia. Since they crossed the mountains the snow has again fallen with us a foot, making the depth on the summit of the mountains about six feet.

Though the weather has been generally uniform, still we have some days of severe weather. On the 5th of December the thermometer fell as low as 42° fahrenheit below zero, freezing the mercury, and possibly a spirit thermometer would have shown a lower temperature. The rear guard of my escort was left eighteen miles behind us to guard certain stores until they can be transported below; and while there one of the men going out to hunt got lost in the woods and mountains, and remained out four days and nights without food or blankets; he became frosted by the cold and became almost deranged, but some instinct led him towards the camp, which he reached crawling on the ground; he was immediately placed under treatment, but his feet were so badly frozen that they became worse day by day; finally, mortification taking place, there was nothing left but to amputate his limbs; to save his life and body the physcian cut off one leg, and the probability is that he will lose the other, if not his life. January 7, both legs have been amputated. Many cases of frost bite have occurred where men were much exposed, but also yielded readily to medical treatment. Many of the cases arose because the men were not well provided with boots, a contingency that I foresaw when starting, and which I tried to meet by purchasing from the clothing department 100 pairs of boots, which I issued to the men, but the wear and tear in the woods and the constant walking to and fro in all weather was such as to take up the whole supply.

The men otherwise are generally healthy, and, though our prospects day by day are snow-clad and dreary mountains, are all cheerful, and long for the opening of spring. In order that the department may fully understand in detail the condition of things and the absolute necessity of a continuation of our appropriation, I have deemed it best to send Mr. W. W. Johnson as a special messenger to the department, who has been with the expedition since we started for the settlements, and who, as a civil engineer, has done some good service with us, and the only man who could make the trip on snow shoes into Walla-Walla. I have instructed Mr. Johnson to return *via* St. Louis and the Missouri river, and meet me at Fort Benton. There are several instruments that I shall need, and a list of which I herewith enclose, and request that they be ordered from Green's, in New York, and be carefully forwarded to St. Louis to meet me at Fort Benton. The command is now to make a more thorough examination of the heights on my return to Walla-Walla in the autumn of 1860. I shall not be enabled to send another express to Walla-Walla before the middle of March, and this cannot reach Washington before the 1st of June, and I could not get a reply before the 1st of August, at which time I shall be at Fort Benton. Therefore, I will submit to the department my plan of operation in the opening of the spring, to be modified as contingencies shall arise; and I also submit certain recommendations, which I now earnestly urge before the department, relying fully upon the fact that they can be necessarily carried out. I shall proceed, myself, with a small party, to Fort Benton, on the 15th of February or 1st of March, locating the road as may be best, and at the same time bring over a portion of my supplies, now there, for my party. I shall push the work with vigor till the 1st of June, when I may send forward a number of men whose services I can best spare, to Fort Benton, to take passage on the steamer. I shall push on the road with the remainder, reaching there by the 1st, or some time in August. I have already at Fort Benton a ninety-foot keel-boat, which I shall use in sending my party down the Missouri.

When reaching Fort Benton I shall send down the remainder of the men, together with a number of the escort, whose times of enlistments will have expired, or near expiration, as an escort to those men as far as Fort Leavenworth. These will be under an officer of the escort. This will not only be necessary, but in the end more economical than to discharge them on the Pacific. If the department does not carry out the military movement I have recommended, then I shall endeavor to dispose of as much of my outfit as possible to the Indian department. The animals can be driven back; but it would not pay to take the wagons back empty, and hence I shall leave them at Fort Benton rather than pay the hire and subsistence of teamsters to take them back. I shall then return with the escort and a very small party back to Walla-Walla, on which trip we can correct many locations made, as well as adding new facts to the topographical knowledge of the country, and at the same time have a sufficient force to repair the road.

But that which my mind has dwelt much upon, and which I submit to the department for a special consideration, is the following, and regarding the decision of which I ask to be informed at the earliest dates, in order that I may provide for all contingencies that the movement must necessarily call forth.

We shall reach Fort Benton with our road, say in August, with a large and empty train, which has cost much. We shall then be compelled either to take this train back empty to Walla-Walla, hiring teamsters for the purpose, or abandon it at Fort Benton. Strict economy, and a desire to see the route practically opened by a new line for military operations, therefore, prompt me to recommend to the department as follows: That a detachment of three hundred recruits, or the number that may be needed to fill up the regiments now in the department of Oregon, be sent, under the charge of a judicious officer, from St. Louis to Fort Benton, in the American Fur Company's steamer, in April or May, to meet our train at the same point, in or about the time of our arrival. Let them start with four months' supplies from St. Louis—sixty days to be used up the Missouri to Fort Benton, and sixty days on the march to Walla Walla. Making use of our empty transportations at Fort Benton, we can immediately return to Walla-Walla in safety and security. I would therefore respectfully ask that, if these recruits are sent, they be turned over to my command at Fort Benton by the officer in command, and I will guarantee to guide and take them to Walla-Walla in security and with success, from which point they can be distributed by the department commander as the requirements of the service may at the time demand. On the score of economy alone I will submit to the department the following figures for its consideration and comparison:

300 recruits from New York city to the Columbia river, at the prices now paid, as obtained from books of Quartermaster General's office, Washington, $200 each	$60,000
Passage of 300 recruits from St. Louis to Fort Benton, at $30 each, (price agreed to by Mr. C. P. Chouteau, of St. Louis)	9,000
Cost of transporting 36,000 rations, weighing, say, 108,000 pounds, at the maximum cost prices per pound, 10 cents..	10,800
Purchases of horses, &c., say	1,200
Hire for forty teamsters for two months, at $50 per month..	4,000
	25,000
Contingencies and rates of teamsters	5,000
	30,000

Saving in cost alone of $30,000, to say nothing of the additional efficiency of the men, the effect upon the Indian mind, and the want of the experiment of testing the value of a new line *via* the Columbia and Missouri rivers, with a land portage of only 600 miles. During the coming summer this can be more successfully carried out than at any other time, for we have at hand the land transportation ready at Fort Benton, and thus have the advantage of comparison of a line within the limits of our territory by water, mostly with one made with all the discomforts to troops, especially *via* the Panama route, through foreign territory.

This will not be to initiate a new movement, but simply a new method or route by which to transport recruits that must be sent out by some route to fill up the regiments now in the department of Oregon.

I am sanguine to believe that if these men start under a discreet and efficient officer, and well clad, and with good shoes, and the number of rations as recommended, they can make the trip in security and safety; and I trust the department will pardon my asking that the command be turned over to me at Fort Benton. Being only guided in this from the fact that my long connexion with the route, and having successfully opened it, and my present knowledge of the details along the line are of such a nature to constitute a sufficient argument for the request, and feeling a special interest in the success of my recommendation. I can carry out the movement without being trammelled by one who is not familiar with the route and the general line of operations between these two points. I trust, therefore, the department will give the matter their special consideration; and whatever the decision may be, I trust to be informed of it at the earliest day.

The Indians on this route are still friendly, and at present we have no special fears for the future. I send by this mail, together with the maps completed, certain reports of the engineers along the line relative to such railroad statistics as have been made out thus far. A special examination and study was given the line of the Columbia by its south bank, (or left,) by Mr. Howard and Mr. Delany, and their reports I submit herewith, together with the line of the road as constructed along the Cœur d'Alène section, and from the Mission to "Cantonment Jordan," together with such of the country at and near the divide of the Bitter Root mountains, and at Sohon's Pass, as we could complete before the mail left. You will observe also, from the accompanying profile of the road from the Mission to this point, the great difference of level between the valleys of the St. Regis Borgia and that of the Cœur d'Alène; and from the accompanying map it would seem that the best tunnel line for the railroad might be by "Sohon's Pass" direct, but this will involve heavy grades on both sides of the mountains for several miles when the grades become easier and more uniform. But I deem a more thorough examination of this special locality needful, as I am sanguine to believe that we shall for the railroad line yet discover a better pass.—(Side map, or line marked A B.) But taking this as it is with a two-mile tunnel, we shall obtain 80 or 90 feet grades, and steeper grades by adopting a shorter or a mile tunnel. The question, then, for our present data will be to use either steep grades and short tunnel, or more moderate grades and long tunnel, or make a more thorough examination with a view of lessening both; and the last idea is *absolutely* necessary in advance of any location. That the directions on an east and west line of the two valleys is remarkably favorable for a railroad location needs no augument. There will be, as might be well anticipated, severe work both of grades and curves, but none so great that will not yield to the skill of the practical engineer.

The line, or so much of it as lies along the Columbia from the Dalles to the mouth of the Pelouse, is entered into with much detail, and the basis upon which rests its peculiar practicability is given at length. So far as the railroad is concerned, there is a break between the mouth of the Pelouse and the Cœur d'Alène lake, because the line of the wagon road cannot, in the nature of the country, be the railroad line.' My own views of the line would be to carry it up the valley of the Pelouse to the Oray-tay-ouse, and thence take it northward to the Spokane, thence along its right bank to the Cœur d'Alène lake. We made use of steamer navigation on the lake (water fifty-five feet deep) and river (water thirty-seven feet deep) up to the point marked "Ten Mile Prairie," ten miles east of the Mission. There the mountain line begins. From a detailed and special examination, made mile by mile, we found that the nature of the ground was such as easily to admit of a line of moderate grades and moderate curves, with possibly short tunnels, through the ends of the spurs. The difference of level between the valley of the two rivers here form the principal problem of difficulties, and yet this is not so great but that it will readily yield to the skill of location. I have had the St. Regis Borgia examined to its mouth, and though it proved too expensive for a wagon road, it is still the line of location for the railroad.

I will not here enter into more details regarding the line in our advance. Suffice it to say that we have run the level line for sixty miles east of the Mission, and found, both by barometer and level, the height of divide 5,100 feet above the sea and 2,804 feet above the Mission. The map has been made in curves deduced from the level, aneroid, and barometer. This is hardly the place to enter into detail regarding all the data that form the basis of the conclusions of practicability, but I content myself with the indorsement of the reports now being prepared by the engineers of the route, who have had experience on the Blue Ridge and Baltimore and Ohio railroads, and in the Alleghany mountains, to say that they regard the line eminently practicable, and as yet have met with no difficulty as great as those met on the latter-named road. I will therefore prefer to prepare and arrange all the details to send the department for their own consideration. I send also, by this mail, an itinerary of the wagon road route from Walla-Walla to Fort Benton, as prepared for traveling; that from Walla-Walla to this point being from our actual line of travel, and the remainder being from our former surveys over the same line. The snow with us at present is three feet deep, though the weather is mild and pleasant. The Indians tell me that winter usually breaks up the last of February, in which case we shall endeavor to resume work in March.

We have received no mail since the mail of the 20th of September, and shall be compelled to resort to snow-shoes till winter is over. I made provision for oats, &c., to be prepared in case I could use horses, but this is impracticable. But we shall be enabled to determine with accuracy the depths of snow and the intensity of cold along the line in midwinter.

On the return of our express men I shall endeavor to send another in April. When the cold weather set in I was compelled to kill the most of my beeves to save them, and shall, I think, with economy and a proper administration, be all safe on the question of supplies.

In a word, the department may rely, though our condition is not such as I could well wish it, still it could be worse.

As Mr. Johnson goes to Walla-Walla on snow-shoes with the expressmen, and will from there push on with all speed to San Francisco, and thence to Washington, he will be enabled to give the department in *detail* all that I might write, and give, indeed, all the details regarding the actual condition of things up to date.

Regarding such instruments and other things, (a list of which I also enclose,) they could all be turned over to the care of Mr. Johnson, to be by him delivered to me at Fort Benton.

I am, sir, with esteem, your obedient servant,

JOHN MULLAN,
First Lieut. 2d Artillery, in charge of Military Road.

Capt. A. A. HUMPHREYS, *U. S. Top. Eng'rs,*
In charge of Office of Explorations and Surveys.

X.

MILITARY ROAD EXPEDITION, CANTONMENT JORDAN,
Bitter Root Mountains, W. T., January 17, 1860.

SIR: Since my communication of the 5th instant nothing has transpired of note, except the arrival of our express from the Bitter Root valley on the 15th, and one from Walla-Walla by three Indians on the 16th, the latter coming *via* Clark's Fork, with horses as far as the crossing of the Bitter Root river. They were a month on the way. They report two feet of snow, the greatest depth met at any point. The expressman from Fort Owen notified me of Mr. Engle's safe arrival in that valley, fears having been there entertained of his loss and that of his party. I enclose his letter, which is all the information I have from him. I daily expect him at my camp; when he arrives, I shall direct him to make his detailed map and report. His letter shows that the condition of his animals rendered it necessary to incur expenses not contemplated. With our present reduced means, I regret to have to report additional expenses.

Our two expressmen, Toohill and Young, are each receiving $100 per month for winter service on snow-shoes, and these three Indians with letters will cost me about $180. Thus do contingencies arise that it is difficult to foresee, and these have to be met, or to be cut off from mails all winter; and I regard the interest of our position too great to let either risk, danger, or expense be considered, when it is necessary to send intelligence to the settlements, the nearest being two hundred and eighty-eight miles.

We are still pushing on the office-work, and hope by the opening of spring that we shall haveevery thing worked up, and be ready for new material that shall enable us to compile a complete map of this region.

The snow with us is still twenty inches deep, but I shall have the men on the road as soon as the ground will enable us to work it; at present it is frozen.

As I informed the department, I shall go to the Bitter Root valley, between the 15th February and 1st March, to prepare a pack-train to go to Fort Benton for my supplies and to prepare for the location of the road for spring's work. Mr. Johnson will have reached the department ere this with my several letters.

Intelligence from Fort Benton and Colville, equally distant, both go to show that the unprecedentedly early winter set in with marked severity, and, as the Indians to-day tell me, by two weeks earlier than they have ever known it.

I am, sir, very respectfully, your obedient servant,

JOHN MULLAN,

First Lieut. 2d Artillery, in charge of Military Road.

Captain A. A. HUMPHREYS, *U. S. Top. Eng'rs,*
In charge of Office of Explorations and Surveys.

XI.

FORT OWEN,
Bitter Root Valley, W. T., March 10, 1860.

SIR: My last communication to the department, of 17th of January, gave you all information up to that date. Mr. Engle was then engaged in preparing our maps and sub-reports. By the 20th of February all the maps were completed, and, with the exception of Mr. Engle's, all reports finished, copies of which are herewith enclosed. Copies of my instructions to chiefs of parties on the various sections of the road are herewith also enclosed, and which will show in detail my programme of spring and summer work. My expressmen reached my camp on February 19, on snow-shoes, bringing dates up to 20th December, 1859, from the department, the latest that I have yet received or can receive before 15th May next. The report of the expressman on the depth of snow, cold, and condition of the road in midwinter is also enclosed.

The winter, after 10th of January, was comparatively mild and pleasant; and though I put the men at work on the road, carrying supplies from the Cantonment to the Bitter Root crossing by hand, a distance of fifteen miles, I shall not be able to work that section of the road till late in April on account of snow, depth being three feet. When the boats are constructed, I shall work with ease up the Bitter Root till the time the grass along the river shall be sufficiently high and nutritious for the grazing of animals to move the working parties.

The winter was most marked in its character; it began early, precipitately and severe, and, though long, ended mildly and moderately, continuing from 2d November to 20th February. It has proved most disastrous to our animals, which, being thin and reduced when winter set in, were not in a condition to reach the Bitter Root valley in safety through the snow and cold, when grass and browsing were covered up.

The grass had been all burnt by the Indians along the Bitter Root river when we reached the Bitter Root crossing, and to reach the main Bitter Root valley, with the low range of thermometer forty-two degrees and snow two feet deep, was a physical impossibility; and I can only regret that the contingency was of a character that human effort could not foresee, meet, or remedy, and hence I have lost many animals—how many I cannot state definitely, as I yet entertain the hope of finding some that may possibly have survived. I have, however, enough left (that succeeded in getting through to the valley) to move my train, as I have thirty-seven oxen and twelve mules and nine horses left, including those of Mr. Engle's purchase. As my train will be light, it shall be small, and we shall work the road with rigid economy. The escort was signally unfortunate with its animals; but it will have animals to move by May or June, when the road is opened, as I send an express to Salt Lake merely to have certain animals (mules) transferred to our service. At Salt Lake, only fifteen days' journey, the quartermaster's department is selling mules for from $15 to $50, and here they are worth $150; and hence economy and other considerations have dictated my course in sending to General Johnson an express asking that fifty mules be turned over to the escort, if he has them to spare. I trust, however, captain, that none of the foregoing circumstances will militate against any movement of troops contemplated from St. Louis to Walla-Walla *via* Fort Benton *this season*. For I can meet all contingencies of such a movement if it takes place; and the only modification that I make to any former recommendation will be that the quartermaster's department, *in case the movement takes place*, be directed to send with the troops to Fort Benton one hundred and fifty (first-class) pack-saddles, with blankets, lash ropes, lash straps, and packing rigging entire and complete. Supplied with this, I can forward to Walla-Walla any number of men under three hundred safely, successfully, economically, and expeditiously. If the department is, then, disposed to act in the matter, it has all the data necessary upon which to base a conclusion.

I left Cantonment Jordan February 26, and reached this point March 5, in order to carry out the object referred to in the accompanying letter to Lieutenant White. I shall return by the 20th instant. Mr. Engle, not being able to reach my camp during the winter, wintered at Fort Owen. He has completed his reports and his map to Fort Benton, and is now engaged in his map for crossing of Snake river, as far as the line was surveyed by him. I shall organize here a pack-train for my supplies at Fort Benton, which will leave in April.

I send a tracing of the map of the line, as completed this winter, by Mr. Kollecki, which has been very carefully mapped, and upon which the approximate railroad line is located. To this I invite your careful and scrutinizing attention, as I believe it is a practicable solution of the Bitter Root mountain tunnel and the location of the road over the section of chiefest difficulty. Mr. Howard has given it much study and attention, and the map and report accompanying it will explain themselves.

My plan of summer's work is set forth in accompanying letters, which will enable us to map in accurate detail a line that I can but regard of marked value at this time.

The Indians are far from being quiet, and though we may have no difficulty whatever, the elements of trouble are evident. The Blackfeet are in open war. How things may turn I cannot say, but we are prepared, I trust, for all emergencies, and prudence shall guide all our movements.

Since passing over the Bitter Root river section of the line I can see that seven years ago we *much underestimated* the work, and even now I shall not give even probable estimates of progress. I can only say that I hope to reach Hell-Gate in August, and Fort Benton in September, and such disposition of the party shall be then made as shall be considered the most judicious. It is also probable that we shall take the line of "Brown's Cut-off," as shown on the map sent in January last. I trust ere this that the recommendation of the department has been indorsed by Congress, and that our renewed appropriation has passed; and, the department knowing the condition of things, by midsummer I would respectfully ask that an amount *not less* than $25,000 (twenty-five thousand dollars) of the new appropriation be placed to my credit *in New York*, and that I be informed of the same at the earliest day. I would also ask that the department forward me, *via* St. Louis, by the American Fur Company's boat, another level for our mountain work.

In case the troops come out, I shall be at Fort Benton in person to meet them, and can direct all details, &c.

I shall write the department again in April, *via* Walla-Walla.

I am, sir, respectfully, your obedient servant,

JOHN MULLAN,

First Lieutenant 2d Artillery, in charge of Military Road.

Captain A. A. HUMPHREYS, *U. S. Top. Engineers,*
In charge of Office of Explorations and Surveys.

XII.

FORT OWEN,
Bitter Root Valley, W. T., March 11, 1860.

SIR: I have the honor to state that I have here secured the services of the friendly Flathead Indians, who leave here on the 15th instant with Mr. Sohon for Fort Benton, with a pack-train of eighty-five horses, for supplies now at Fort Benton. They will rejoin me by the last of April. In addition to the economy and expedition of this arrangement, much good is anticipated from it, in arranging the unsettled condition of affairs with the Blackfeet Indians, with whom disturbances have again commenced. I write to the Indian agent, Colonel Vaughn, at Fort Benton, to send the chief "Little Dog" and some of the principal men to meet me on a visit, in order that I may assure their tribe that we come as friends, and in order to guarantee our safe conduct while working the road through their country. Unfortunately for the country, there are numerous ill-disposed whites and half-breeds among them, who, by a studied course of evil design, are daily instilling into their minds things well calculated to produce the most pernicious results. The Indians hence are much exercised, and being in open war with the Pend d'Oreille Indians, are disposed to be troublesome. I shall, however, by a judicious course, assure them of our good intentions, and am sanguine to believe that the many conflicting elements may be so governed that much good may yet accrue from our mission.

The Indian affairs in this quarter are *not* in the condition that I would well wish them, and I fear, unless measures, speedy and remedial, are taken, that we shall have much trouble ere long in this region. The Blackfeet are, I believe, if not meddled with, a good people, and might be made better; but the evil elements are so directed that it is evident that it is only a question of time that they shall be included with other miserable and treacherous tribes. But I hope for the best.

I would state that during the past winter, when the snow had overtaken us, many of the settlers of this valley came forward with their animals to our relief, and though their services were small, their privations were many, and in midwinter.

I have told them all that I will allow them a fair compensation for the services of themselves and their animals, but that I can only pay them out of the renewed appropriation. I shall make a statement of the same on my accounts for the month of March. I return to the cantonment on the 15th of March, having completed satisfactorily the objects of my mission to this point. Trusting that the department, appreciating the contingencies of our past condition and the possible events of the future, will be disposed to indorse my views and recommendations,

I remain, truly, your obedient servant,

JOHN MULLAN,
First Lieutenant 2d Artillery, in charge of Military Road.

Captain A. A. HUMPHREYS, *U. S. Top. Engineers,*
In charge of Office of Explorations and Surveys.

XIII.

CANTONMENT JORDAN,
Bitter Root Mountains, W. T., April 3, 1860.

SIR: My last communication to the department was of the 10th and 11th of March, dated at Fort Owen, Bitter Root valley. Having completed there the objects of my trip, I returned to this point on the 28th of March. Mr. Sohon, my guide and interpreter, left Hell-Gate on the 16th of March, with a pack-train of eighty-five horses and seventeen Flathead Indians, to bring me eleven thousand rations from Fort Benton from the stores there, sent in the summer of 1859 from St. Louis. I anticipate his return by the 16th of this month. I also visited the Pend d'Oreille Indians to talk with them and assure them of our friendly intentions. I also sent letters by Mr. Sohon to Colonel Vaughn, the Indian agent at Fort Benton, requesting him to send me a delegation of three or nine Blackfeet chiefs, in order that they may see the true objects of our visit, and in order that we may pave the way for a successful passage while working the road through their country. These objects gained, we shall, I trust, have no difficulty in regard to our relations with the Indians.

On returning to the crossing of the Bitter Root river I found the ferry established, six boats constructed for the movement of our parties while working the road along the Bitter Root river, (forage not being in sufficient abundance for our animals,) and the road constructed and completed for ten miles eastward from the crossing of the Bitter Root. The banks of the river graded at the ferry and ford half a mile below, and the road cut by Mr. De Lacy's party for three miles westward from the ferry towards the cantonment, when the snow, still eighteen inches deep, prevented the party from working further. Finding that Mr. De Lacy's party could not work this section of the road, on account of snow, I moved them to a point ten miles up the Bitter Root, where I shall have work for two or three weeks, and where I have now eighty-five men at work with picks and shovels. At this point a bald prairie spur makes down to the river's edge, and there is no way of passing it, except at the cost of heavy grading. This once completed, we shall, I trust, have the road completed for twenty-five miles up the Bitter Root by the 1st of May. Reaching the end of that distance I think I shall leave the immediate line of the Bitter Root river and take the line of "Brown's Cut-off," as laid down on the map sent to the department in January last. I endeavored to examine this in detail on my return from the Bitter Root valley, but the depth of the snow precluded the possibility of a satisfactory examination. I am sanguine to believe, however, it will prove to be our best approach to the Hell-Gate defile. As soon as I despatch the express for Walla-Walla, I shall go in advance (about the 7th of April,) and, with the guide I have hired for the purpose, make a special examination of the line, and my next letter to the department will contain a detailed map and report on the same. The work ahead is progressing satisfactorily, and the men, after a long, dreary, and cold winter, are as anxious as I am to reach the Missouri. On account of the snow, eighteen inches to two feet deep still, I am unable to work the section of the road from this point to the Bitter Root ferry; but by the last of April I hope to have the men on the road. At this point the timber is so exceedingly dense that the light of the sun cannot penetrate it to thaw the snow; whereas along the Bitter Root, where the timber is open, and with a southern exposure, the snow has already left the ground. Were it not for the dense, thick timber in this valley, the snow would be off the ground by the 1st of April, but owing to the thickly shaded forest it cannot be off till 1st of May. Thus has the dense timber proved our most serious difficulty to contend with, both winter and summer, while in the Bitter Root mountains. From here to the ferry there are ten miles of heavy timber-work and grading to be done, involving the crossing of a mountain spur 1,200 feet high, where the continuous grading from base to base will be one mile, mountain sloping at an angle of 50°. From the character of the work I have estimated that the parties in advance will open the line to the Hell-Gate defile by the same time that we can open the line hence to the Bitter Root ferry. Should I reach Hell-Gate by the 1st of August I shall rest satisfied. This will give me to the 15th of September to work the road to Fort Benton, which will be ample for our purposes, both for those who return to Walla-Walla and those who will descend the Missouri.

I have moved our observatory, under charge of Mr. Weisner, from the Bitter Root crossing to the Hell Gate defile, at the entrance, where it will be established for one or two lunations, and I shall then move it to the summit of the Rocky mountains, at the head of either the Little or Big Blackfoot Pass, where, remaining one or two lunations, I shall move it to Fort Benton, in order to get there a good series of observations, including the eclipse of 17th of July next. We have five months' transit observations at the Bitter Root ferry, and the work thus far has been very satisfactory. Our transit, which is one that Lieutenant Jones had on the Colorado expedition, and our chronometer, are all in good order and work well.

I found the party under Mr. Kollecki progressing rapidly with their survey from the Bitter Root ferry eastward, and every detail of a topographical character is noted down.

Mr. Conway R. Howard, civil engineer, accompanied me to Fort Owen, intending to go with Mr. Sohon to Fort Benton to examine the heads of the Little and Big Blackfoot passes for wagon and railroad practicability; but on reaching Fort Owen his old complaint of a swelled groin attacked him, and he was unable either to go or to return with me, and I was compelled to leave him at Fort Owen.

Mr. Engle, not having completed his map of his summer's field-work, was also left there, and the cook, Crirarre, to cook for them. My wagonmaster, Mr. Caldwell, accompanied Mr. Sohon to Fort Benton to assist in bringing supplies over.

My animals are at the Hell-Gate ronde, in good grass, under charge of Van Dorn, my principal herder at the cantonment.

Everything is ready for our movement eastward as soon as the road shall be opened, the wagons, tools, &c., being put in complete order during the winter. The men, with the exception of four soldiers sick in the hospital, are healthy. I regret very much to report that the scurvy made its appearance among the soldiers on the 10th of March, but by a timely supply of fresh succulent vegetables, purchased from the Pend d'Oreille Mission, the disease was abated, and all traces have disappeared. The most remarkable phase of the malady, however, was, that while some twenty cases occurred among the *soldiers, not one* occurred among the citizens. The reason of this was plain, when the department learns that the kind and quality of the rations given the men were exactly the same, except in the case of the desiccated vegetable. I issued to the citizens double the quantity of desiccated vegetables that Lieutenant White, the commander of the escort, issued to the soldiers, the amount allowed the latter being that laid down and issued according to the regulations of 1857. The bean, rice, and vegetable ration form a unit. I issued half beans and half vegetables to form the unit. The regulations say half beans, quarter rice, quarter vegetables for the soldier; but for the citizen, who works harder, longer, and more than the soldier, the allowance was modified to suit our special case, the unit still remaining the same. I enclose the report of the physician in this special case, as I deem it one worthy to be brought to the special attention of the War Department. The desiccated vegetable has been but recently introduced into our service, and its *true virtues* as an antiscorbutic have not possibly been so rapidly tested as in our present case, which proves that it is an antiscorbutic, provided the quantity be great enough. Our men have been so situated that they could not procure, at any place or cost, fresh vegetables, which case rarely occurs but with the army at the present day; but the snow and cold, acting as secondary provocating causes, doubtless led to it. I would therefore, based upon the experience of the past winter, recommend that where any portion of the army is situated as we have been during the late severe winter the amount of desiccated vegetables be increased, if not doubled. It would be well to ascertain the facts in the case of the army at Fort Bridger, in 1857, provided the desiccated vegetable was used there. But the winter there was not as long or as severe as with us, and the men generally were not as much exposed. When the scurvy attacked Dr. Kane's expedition in the Arctic regions, it occurred on the opening of spring, when, though he was provided with antiscorbutic, his men were exposed to severe cold and hardship.

I am glad to report that this dreaded malady has disappeared, and the men cheerful and strong. For the timely procuring the fresh vegetables—the surest, if not the only remedy—I am under obligations to the Jesuit fathers of the Pend d'Oreille Mission, who let me have them at the expense of depriving themselves.

I also enclose the several reports of Mr. Engle on his several trips of reconnoissance during the past summer, which, being written at Fort Owen, I had not an opportunity to forward them at an earlier date. They are full as to detail, and show in a truthful manner the leading characteristics of the routes examined. Taking our main line of travel and work as the basis of operations, these several reconnoissances show the side work, and, with the topographical notes now being worked up, will, I trust, put the map of the country in a somewhat full and reliable light. I have sent Mr. De Lacy, with a small party, to make a survey and report upon the Bitter Root river from the ferry to its junction with the Clark's Fork, and up the Clark's Fork to the Jocko river, and to the Pend d'Oreille Mission, when, connecting the latter with the Hell-Gate defile, it will enable us to map the country in a satisfactory manner. This *line has never* been correctly mapped. Governor Stevens sent Mr. Lander to examine and report upon it, but he left no reliable data of his work; and hence, in order to show not only this water course from source to mouth in its true line, we shall also map the country between the Clark's Fork and the Bitter Root. Mr. De Lacy starts to-day. Lieutenant Howard, of the army, is in military charge of the soldiers in advance, and from his co-operation I look forward to the best results during the coming summer.

Our wagons, property, &c., under charge of a portion of the escort, and commanded by an officer of the escort, will remain at this point until such a time as the road shall be opened hence to the Hell Gate, when we shall take up our line of march eastward; and I can only hope that three month hence the department shall be informed that we shall have left another hundred miles behind us.

Any recommendation or suggestion regarding a movement of troops, or regarding our appropriation coming now too late to be of any avail or effect, I refrain from making them, but still am sanguine to believe that Congress will be disposed to so act that our special wants may be provided for.

To abandon the work when work is most required, and at its maximum, might be in accordance with the possible action of Congress, but certainly not in accordance with my ambition and desire to see the route practically and successfully opened, nor can it be in accordance with the spirit that explored, projected, planned, and has, thus far, constructed it. And though the non-action of Congress may have the effect to lay my plan of progressing (even without means) open to censure, still I feel that public necessity and public economy dictate that our work should be pushed on to completion with the force now working it.

My express will return by 1st of May, bringing dates from the department to 5th March. I shall immediately despatch it on return to Walla-Walla, reaching there 16th May, bringing dates to 5th April. As soon as the high water permits, I shall have monthly expresses to Walla-Walla, and shall keep the department informed of our progress and movements.

The snow at the cantonment is now eleven inches deep, but in the timber hence to the crossing of the Bitter Root river, where the road is to be located, the snow is still eighteen inches to two feet deep.

My mail will still reach me *via* Walla-Walla.

I am, sir, respectfully, your obedient servant,

JOHN MULLAN,

First Lieutenant 2d Artillery, in charge of Military Road.

Captain A. A. HUMPHREYS, *U. S. Top Engineers,*
In charge of Office Explorations and Surveys, War Department.

XIV.

MILITARY ROAD EXPEDITION, ETC.,
Forty miles east of Bitter Root Ferry,
Camp in Bitter Root River, W. T., June 6, 1860.

SIR: My last communication to the department was under date of April 3, 1860, at Cantonment Jordan. Having completed all my arrangements at the last-mentioned point, I joined the advanced working parties, then working a spur of the mountains about ten miles east of the crossing. This was the first difficult point met with, and upon which with a force of eighty men we were engaged about two weeks, cutting for three-fourths of a mile a road along the spur of the mountains, in places the grading being fourteen feet deep; a portion of this was difficult on account of rock, which required blasting. From this point to a point marked "Brown's Cut-off," which was marked on the map sent the department last winter, 25 miles east of the ferry, our road lay over timber plateaus, the work consisting in timber cutting and clearing and the grading of hills and the points of spurs of the mountains that made down from the river. From the Bitter Root ferry to the first most difficult point the work consisted in cutting through open timber, and grading at points where the road passes from one plateau to another; the formation along the Bitter Root on both banks being a series of plateaus of different levels. By the 20th of April the road was opened from the ferry to "Brown's Cut-off," 25 miles. From that point onward to Hell-Gate I was still undecided relative to the location of the road, as I wrote to the department three routes presented themselves for examination: one following the right bank of the Bitter Root; a second involving two additional crossings of the Bitter Root, in order to avoid these difficult and rocky mountain spurs; and the third by "Brown's Cut-off." This last I have heard much of from Indians and others, and from all the representations made of it I could only conclude that it was *eminently* practicable for our purposes. On the 20th of April, therefore, with a small party, I started to explore and examine it; the first section of some eight miles up and near the "Nemote creek" (see sketch) I found very good; for the end of this distance, however, we entered a mountain gorge or narrow defile, where our progress was impeded by rocks and fallen timber, which, however, proved not the least of our difficulties. After clambering over this exceedingly difficult region for three days we encamped at the foot of the divide of the mountains, with snow-covered mountains in view from every point. On the fourth day we started to make the passage of the mountains, when within two miles of the summit our progress was stopped by three feet of snow, rendering it with the dense timber impossible for us to travel either on foot or on horseback. I had enough information, however, to convince me of its entire impracticability, and sick at heart and fatigued with our exertions we were compelled to retrace our steps back to the Bitter Root. I saw that, even had the topographical features of this section been at all favorable, the great depth of snow would at all times form an insuperable objection to the location of any road by that route. I had, therefore, but two other locations to choose between. The one to follow the right bank of the Bitter Root with all its excessive and difficult work in rock, earth, and timber, to grade a road of five miles in length on the mountain side, or avoid this by a shorter and easier line, but involving two additional crossings of the river; neither of which crossings could be well suited to safe ferries, owing to the rapid water and high, rocky, almost precipitous banks. Truly the matter was not of easy solution. On the one hand, the great length of time (apparently) to work the road, our scanty means, the putting in jeopardy the opening of the remainder of the line this season to Fort Benton, the probable necessity of the line being needed for travel this season, the question of supplies, &c.; while on the other, the disadvantages in our position, the difficulties of additional ferries, the greater inconvenience to the travelling public, the increased taxes of tolls, the liability to have the ferries destroyed by Indians or swept away by freshets, and the delays to travel naturally incident to frequent river crossings. These and many other minor elements all tend to render my mind uncertain relative to the most judicious course to pursue. The engineers of the party were in favor of additional ferries, saying "that five months would be required to work the five miles around the mountains," fearing from the exterior appearance of the mountains we should have solid rock nearly the whole distance, thus compelling us to use the slow and tedious process of blasting. By the first day of May the road was completed to the western foot of this mountain, a distance of 36 miles east of the ferry. With all the facts before me relative to the location, I determined to avoid the crossings and to keep the road on the right bank, and on the first day of May we began work on the western side of the mountains with a force of 65 men—this being the greatest force that I could spare for this section of the line, for on the same day I despatched 25 men to the Bitter Root ferry, under Captain W. W. Delany, to open the line from the crossing westward to the mountains towards the cantonment; this, owing to snow, we could not work at an earlier date, and hence we left it untouched until such a time as the advanced spring should cause the snow to disappear.. On the 20th of April Lieutenant White, 3d artillery, with a force of 32 men, began work also from the cantonment to the mountains eastward to connect with Captain Delany. This line, 15 miles in length, led through some of the most dense timber we have had in the line, and involved a mile and a quarter of grading over the spur of the mountains on the two sides. Rainy weather was much against them; but I am gratified to inform the department that by the last of May the work was completed, and that our trains of wagons have already passed over it. On the 1st of June all the escort wagons made the passage of the Bitter Root ferry in safety, and on the 4th day of June my wagons started, and the whole train is now encamped at the Bitter Root ferry, 70 miles from Hell-Gate.

We therefore abandoned our winter home on the 4th of June, where the fetters of winter and our difficult work ahead of us had bound us for six long dreary months. The advantages of our being enabled to get below all the crossings of the St. Regis Borgia to find a place for winter are more marked now than ever; for all the mountain streams are now torrents and swimming, rendering it impossible without suitable bridges to cross them. But being *below* all the crossings our work hence to Hell-Gate will not be interfered with on account of high water, and I trust by the time we reach the Hell-Gate all the streams will be fordable. Early in April I sent Captain Delany, with two men, down the Bitter Root in a boat, and thence up the Clark's Fork to the mouth of the Jocko, and thence to the Pend d'Oreille Mission, in order to explore in detail that particular section hitherto *unexplored*, and report upon its adaptability for wagon and railroad purposes and its general characteristics. We made two trips and returned safely by last of April.

His report is herewith enclosed. I also enclose the last of Mr. Engle's reports, which, together with the map of the line now compiled, gives much topographical information hitherto not given in as accurate and detailed a manner as has often been desirable. When I last wrote the department, Mr. Sohon was absent with a band of Flathead Indians, in order to bring me stores from Fort Benton. He returned the middle of April all safely without any loss.

The delegation of Blackfeet Indians, under the principal chief "Little Dog," accompanied him in order to be assured of the *true* objects of our mission. They were well received, treated kindly, and, receiving a few presents, returned much pleased to their people. Certain malicious white men had, by false reports, spread quite a panic among them, and they had augured anything but favorable objects in our approach to their country. I am confident that by a prudent and judicious course with them our passage will be fraught with neither delay nor difficulty.

Knowing that on our arrival at the Hell-Gate ronde our principal difficulty will be at an end, and the road hence to Fort Benton easy of construction; and having wagons and many articles that we should not need on the road, and which in the course of time will be more available at that point than elsewhere, I directed a storehouse, corral, and depot to be established at a central and notable place in the Hell-Gate ronde, and which I have called "Camp Humphreys," where such things as we do not now need will be left under lock and key, and where the presence of the settlers living in the valley will be a sufficient protection to guard them for the present. Our observatory, under Mr. Weisner, was established at Hell Gate until 10th of May, where we succeeded in collecting six weeks' good observations. These were left there; tried other fixed points where it would be desirable to collect observations, the divide of the Rocky mountains and Fort Benton.

But being desirous of observing the eclipse of 17th July, it was thought best to move the instruments on the 10th May to Fort Benton at once to observe there till 1st of July, and then return and observe the eclipse and take other observations for the month of July from the summit of the mountains. At that time the atmosphere will be clear and free from haze. At present in the mountains it is hazy and difficult to observe; whereas at Fort Benton the sky and atmophere are clear, enabling us to make good use of the time. Mr. Weisner arrived safely at Fort Benton the 20th of May. He crossed the mountains by "Lewis & Clark's Pass." I had thought that if, with a moderate expenditure of time and means, I could construct the road along the Big Blackfoot valley, it would be not only more direct, but shorten the route some fifty miles. I have never been over the Big Blackfoot myself, and I have been unable to gather from Governor Stevens's report the amount of work that would be required to construct a road. It is true, an itinerary is therein given by Lieutenant Donaldson, but I am assured, with all respect to his memory, that it is far from being reliable, and does not probably include one-fourth of the actual work. With a view of ascertaining the exact character of the line, I directed Mr. Sohon to accompany Mr. Weisner as far as the summit of the mountains, and to thoroughly explore the line with the passes at its head, Lewis & Clark's and Cadotte's, and a third between these two. This examination proves the line—except at great cost and time—unsuited to our present work. For 20 miles from the mouth of the Big Blackfoot the rocky spurs and cañon-like formation of the valley prevent any line except at great cost. This examination of the line, however, will add much topographical information to our present knowledge, which, together with an examination to be made in July by Mr. Kollecki and Mr. C. R. Howard, civil engineer, will clear up our doubts regarding heights, grades, and difficulties regarding the wagon road and railroad location.

In December last I had the section of the St. Regis Borgia cañon from its mouth to Cantonment Jordan examined. It proved impracticable for wagon road purposes, but the trip was made under marked difficulties. Three and four feet of snow and dense timber rendered the exploration exceedingly difficult for the explorers, and they suffered much, one of the men being frosted by the exposure. The topography of the country, however, indicated it as the railroad line; and in order to satisfy ourselves regarding it, and thus make known this *unexplored* link of twelve or fifteen miles, I directed Mr. Kollecki and Mr. Howard to explore and survey it. They started 25th May; but having gone up it some three miles, the freshet compelled them to return. They started from its mouth upward. They then went up to the cantonment to try with a small wagon boat to descend the St. Regis Borgia. They started on 29th May; but before they got far the impetuous current and the difficult character of the stream, rendered so by fallen trees, snags, &c., precluded the possibility of its thorough exploration. We have information enough, however, to convince us that it is the railroad location. The survey of the wagon road and the adjacent country keep pace with our progress, and every detail is collected with much care, which will enable us to show in truthful light the character of the line. The level line is still being run, and the facts thus far collected prove most markedly the erroneous judgments, opinions, and reports heretofore made of this section by Colonel *W. F. Lander* in 1853. The data will, at the proper time, be in the possession of the department. In order that I might, from positive information, know something of the character of the spring freshets, I directed the man—one of our own men—in charge of the ferry at the Bitter Root crossing to construct a water gauge, which has been done, and upon which we have recorded the daily rise since the freshet began, and up to date. The Bitter Root has risen *eight* feet, and, so far as the observations of our party go, there are no signs of a greater rise than *ten* feet. The river reaches its maximum rise by the 10th or 15th June, and then falls rapidly. When it is known therefore that the banks of the Bitter Root are from twenty to forty feet high, the danger of freshets to the wagon or railroad line is either a chimera or a false assertion.

During the winter our expressmen measured the depth of the snow on their several trips in every winter month, and from these we can construct an accurate snow profile from here to Fort Walla-Walla, and thus from actual measurement say how deep *exactly* the snow was at any given point.

This has *never* before been done on any line; and from the fact that both Colonel Lander and Mr. Beall were compelled or rather did suspend work on their lines, respectively, early in autumn, to be resumed only late in the summer, is sufficient proof to my mind that no *winter* reports can be made by these men; for if it had been practicable to winter on their lines, why was it not done?

In connexion with our present operations, I thought when the expedition should have concluded its field labor, that, in discharging my men, I should cause one party to return east by way of Fort Laramie, and thus explore a direct line from Hell-Gate to Fort Laramie. But the loss of animals, scanty means, and the danger of doing it with a small party, have been sufficient to deter me from attempting it at present, and especially now, since I learn that Capt. Reynolds is about to do the same thing.

I am sanguine to believe that he will find from Fort Laramie an excellent line to connect with our line either at the Deer Lodge or at the head of the "Little Blackfoot" river. I reported to Governor Stevens regarding this line in 1854, and should then have made it had not my special instructions precluded it. Several times since I have referred to it in conversations with Governor Stevens, Lieutenant Warren, and others; and I also learned of Colonel Ferris, Indian agent at Fort Laramie, that he had been over much of the line, and represents it as eminently practicable for wagon road purposes. In the Bitter Root valley are living many half-breeds and Indians, several Shawnees and Delawares, and they all assure me that an easy line can be had, and that they are willing to go with any party sufficiently strong to oppose either the Sioux or Crows in case of difficulties, and guide them through it. I know not by what line Captain Reynolds will travel, but should I learn of his whereabouts, I shall despatch two trusty Delaware Indians to him to assist in his expedition and to guide him safely. I feel much interest in the exploration of this line, and for the great saving of distance and excellent country through which the line will pass, am confident that it will become the emigrant route to the north Pacific; and in case of a central railroad line along the Platte, Fort Laramie becomes at once the best and proper point of deflection for the north Pacific, *provided* the link from Fort Laramie to the Hell-Gate proves practicable.

The valleys of "Beaver Head," "Deer Lodge," Hell-Gate, and "Bitter Root" are large and arable, and must at once become centres of population. Indeed, even now they would be such, did not the fears of the Blackfeet, Crows, Sioux, and Snakes render it a physical impossibility.

Mr. Sohon has returned from the Big Blackfoot exploration. He found five feet of snow on the summit, which, together with the cañon-like formation at many points, and in others on account of its rocky character, renders it, with our present time and means, out of the question for the wagon road location; and hence I regard that our original line, *via* the Hell-Gate and Little Blackfoot, will be, even at the expense of distance, the best line to follow. This line, as you know, too, is free from snow in winter, and at all times can be travelled, and, looking towards the connexion with Fort Laramie, the Hell-Gate line is preferable.

Regarding the work now in progress, I would state that, up to date, three of the five miles of earth and rock grading is completed, and, in addition to two miles of earth and rock grading, we have fourteen miles of work to the Hell-Gate. I find that we have much blasting, and at present I have four sets of drills going every day; the rock being a hard, flinty limestone, this has been inevitable. I have at present one hundred and twelve men on the road, and we are pushing through with our greatest exertions to reach Fort Benton early in August next.

The freshet is high this year, and in places gives us much trouble; streams will be fordable, however, early in July.

I have the honor to be, respectfully, your obedient servant,

JOHN MULLAN,
First Lieut. 2d Artillery, in charge of Military Road, &c.

Captain A. A. HUMPHREYS, *U. S. Top. Engineers,*
In charge of Office of Explorations and Surveys.

XV.

MILITARY ROAD EXPEDITION,
Camp at Fort Benton, Nebraska Territory, August 2, 1860.

SIR: I have the honor to report the arrival of my expedition at this point yesterday, the road of 633 miles from Fort Walla-Walla to Fort Benton being opened. I submit the following summary of my operations since the date of my last communication with the department.

By constant and unceasing labor we were enabled to complete the grading and rock blasting around the mountain on the Bitter Root river, referred to in detail in my last report, and on the 20th of June our wagons with our working parties moved over it as far as the Chil-mo-e-pah creek, which we were obliged to bridge as the water was deep and impassable for wagons. This was a heavy structure, being 148 feet in length, with four heavy cribs in the centre; it occupied ten days' work of forty men, and is durable and commodious. Hence to the crossing of the Skah o-tay creek, the road passes for a quarter of a mile over bottom land; then rises to an open timbered plateau, along which it continues for two and a half miles, and then passes through somewhat dense timber for a quarter of a mile to the crossing of the Skah-o-tay. A distance of two miles will be saved by constructing a bridge across this creek at its mouth; but as this would have caused at least a week's delay, we located the road up to the crossing about two miles above its mouth. From this ford the road ascends a hill of four hundred feet in height, and then taking successive open timbered plateaus, crosses the Kul-ko-lau creek, and comes into the western end of the Hell-Gate ronde, a large prairie basin, some twenty miles in length by ten in width. This we regard as the termination of our heavy work. We called our camp here Camp Humphreys, 336 miles from Walla-Walla. Our wagons, with the working party, reached this point on the 23d of June.

I deemed it most prudent to make here a depot of baggage and other property not actually needed for our field duties, and also held an auction for the disposal of such articles as were salable, with a view to disbanding my party and settling everything at Fort Benton, according to instructions received. I transferred to the Indian department such property as was no longer needed, repaired the train, and on the 28th of June resumed our march up the Hell-Gate valley. On the 2d of July we reached the crossing of the Big Blackfoot, a distance of twenty-two miles from Camp Humphreys. We were obliged to do some three hundred yards of grading in the last five miles before reaching the crossings, the road leaving the level prairie and passing over the foot slopes of the mountains. The water being very high, we were obliged to ferry our wagons over, which was accomplished with only one accident; the boat upset, emptying one of the wagons into the river, which gave us some difficulty before we could get it out. Here we had a depot of supplies for our onward movement, a number of men were discharged, and we moved from here on the 3d to the crossing of the Hell-Gate, five miles. An express was sent to Major Blake with a reply to his communication of the 15th of June, received 2d of July. The road to the crossing is over a fine level, open timbered plateau; not much work required. We had to ferry the first crossing of the Hell-Gate, which occupied the entire day of the fourth of July; five more crossings were made in the next twenty-three miles, the valley being narrow (quarter of a mile in width) and the river tortuous in its course. These fords were, at this time, deep; indeed, at the fourth crossing, the oxen of one of the wagons became unmanageable, turned down stream, and the result was the drowning of an ox and a detention of three hours. These fords can be avoided by side hill excavation, part in rock, which time would not allow us to perform. The amount of work required on this portion of the road to render it practicable at all seasons will be one month's work of fifty men.

We were detained two days to cut a graded road over the butte at the Beaver Dam creek, to avoid two impracticable crossings of the river. We continued up the valley of the river, making nine crossings in forty-one miles—in all, fourteen crossings. The distance from the mouth of the Big Blackfoot is sixty-four miles; from this point the valley of the Hell-Gate becomes wider, the timber becomes scattering and thin upon the mountains, and bunch grass covers the valley and hill-sides. After passing mile-post 400, the valley of the river is about a half mile in width, and the mountain slopes become less abrupt; the road now leaves the river and crosses over rolling foot-hills, and through intervening bottoms for eleven miles, to Flint creek; thence twelve miles to Benetzi's creek, it passes over nearly level bottom sand. After leaving Benetzi's creek we made three crossings of the Hell-Gate and two of the Little Blackfoot. These crossings can be avoided by grading and rock blasting, requiring ten days' work of thirty men. From this point the road passes for twenty-one miles over rolling hills by the Deer Lodge prairie, and crosses a high hill (elevation 1,000 feet) to the Little Blackfoot. This portion of the road will be relocated to avoid ascents and descents, and materially improved and shortened. To take the line *via* the Little Blackfoot valley will require five days' work of fifty men. We were obliged to make a crossing and recrossing of the stream, which can be avoided by a half day's work of fifty men. The present location of the line hence up to the foot of the divide will require shifting up on to the foot-hills to avoid soft ground; this would require two days' work of fifty men.

On the 18th of July we made the crossing of the Rocky mountains at Mullan's Pass, distance from Walla-Walla 469.8 miles. The ascent is easy, the descent to the east is steeper, but will not require to double teams. The road was cut for a third of a mile through timber, thence with light grading and clearing along the Big Prickly Pear creek to Fir creek, (476 miles,) where we were obliged to do some two hours' work—chopping, clearing and grading. The road passes hence over rolling prairie for 20.2 miles to the Small Prickly Pear creek. In this distance we crossed six small streams, each of which required a small amount of grading to allow the passage of our wagons; we also did some 300 yards of earth excavation, which, however, delayed the train but little.

We remained three and a half days in the Small Prickly Pear region, being stopped by severe work on the Medicine Rock mountain. As thorough an exploration as we could give the country at this time showed that the only practicable line lay over the mountain, and accordingly we were obliged to do considerable work, principally in earth excavation. After crossing this mountain, the country in advance was still broken and knobby, the river flowing through an impracticable cañon. We followed the trail for ten and a half miles, when the valley becoming wider it was followed down for four and eight-tenths miles, making twenty crossings—the trail being very rocky, and requiring four to six days' work to make it passable by the high water trail. On the 25th of July we reached the last crossing of the Prickly Pear, and also the last point where work of any account will be required.

In reference to the region of the Medicine Rock mountain, I would state that it is truly the most difficult between Hell-Gate and Fort Benton, and compares unfavorably with any section of the same length on the entire road. The ascents and descents are many and great; and though three days were occupied in exploration, the location was not satisfactory. It is believed that a better location will be found by turning to the north, some of the spurs making down to the Prickly Pear cañon. The ground is rocky and broken, rendering excavation difficult. The line as worked was thought best at the time. What is needed is a more thorough examination and additional work. Fifty men for ten or twelve days could be profitably employed on this section.

The road from the Little Prickly Pear creek to the Dearborn river passes for sixteen miles over rolling prairie covered with fine grass. Water is found at a point seven miles from the Prickly Pear.

The Dearborn river is about 120 feet in width. It is always fordable, with the exception of about two weeks in the spring. From the Dearborn to Sun river, 32.5 miles, the road is over nearly level prairie, passing by Bird-tail Rock, a prominent landmark, at the base of which is a good camp for grass and water, but no wood. Distance from the Dearborn, 16.5 miles.

On reaching Sun river, the question arose about the route thence to Fort Benton, the usual line travelled being by the lake and spring, and a point called the Big Coulée, on the Missouri, 11½ miles above the fort; thence down the river to Fort Benton. This involves a march from the spring to the Missouri of 22 miles without water. Wood is very scarce, and the supply of water even is precarious. It was, therefore, determined to explore four lines from Sun river in to Fort Benton. Mr. Sohon, with a small party, was sent *via* the Mission across to the Teton river, with a view of ascertaining the feasibility of a detour from a point near the Bird-tail Rock, striking the Sun river at the Mission, about 10 miles above our crossing; thence passing to the Teton, and thence down to a point nearly opposite Fort Benton, where the line would cross to the Missouri. I proceeded with a second party across to the Teton by a line leaving the Sun river about a mile below the road crossing, while the main train in two parties travelled the usual routes to Fort Benton. One by the route already indicated, *via* the lake, spring, and coulée, while the other proceeded direct from the spring to the fort.

The parties all met at the terminus of the road on the 1st instant; and on comparison of notes, it is thought that the line from the crossing of Sun river, at the Agency farm, *via* the Teton, is the best for wagon trains, especially when drawn by oxen, there being only one camp without wood, and one march of 18 miles without water. The supply of water on this line is always abundant and the grass good; the distance, however, is increased 18 miles, but this is counterbalanced by the advantages of short marches and a good supply of water.

In order to determine the question of the practicability of the line of the Big Blackfoot, Mr. Howard and Mr. Kollecki were sent over it with a small party to make a detailed survey of it, and report upon the amount of work required for a wagon road location, &c. The report has not yet been prepared; as soon as received, it will be forwarded to the department. I can only state, as the result of this survey, that this line presents so much work as to be impracticable for a wagon road except at heavy cost. Mr. Howard will prepare the profile and report at Washington, concerning which I will address you a letter.

On my arrival here I turned over to Major Blake all of my wagon transportation. I now propose to return over the line of the road, according to instructions received from the department under date of May 30. I retain one assistant and a force of about 25 men. We shall move with a pack-train in advance of Major Blake's command, and shall change the location of the road already mentioned in this and my previous reports so far as practicable, and shall spend the remainder of the season in working those points most necessary for continuous communication at all seasons. From our examinations of the Bitter Root river, it will be impossible for us to construct a bridge over it this season. It is 440 feet wide from bank to bank, has an exceedingly rapid current, with a rise of eleven feet in the freshet. The driftwood floating down it in the spring rise would sweep away any structure unless built with stone piers, for which I will forward a detailed estimate; but it would be safe to say that a structure suited to this river crossing would cost not less than $25,000, or one-fourth of our present means for the entire route. An examination will be made of a line from the Cœur d'Alène Mission westward to the north of the Cœur d'Alène lake; thence along the Spokane, crossing it and connecting with the present line of the road at the Camas prairie or Nedlhwaldk creek, with a view of avoiding the flats of the Cœur d'Alène and St. Joseph's rivers, which are impassable for wagons before late in July or early in August. Here a radical change of location will be needed, and a new section of 25 miles from the head of the Spokane to the Mission will have to be worked to get the most favorable permanent location.

As soon as we reach winter quarters I shall prepare and forward a complete and detailed report and estimates for the line of the road, with bridges, &c., complete, where they cannot be avoided.

I am, sir, respectfully, your obedient servant,

JOHN MULLAN,

First Lieut., 2d Art., charge of Mil. Road Expedition.

Captain A. A. HUMPHREYS, *U. S. Top. Eng'rs,*

In charge of Office of Explorations and Surveys.

XVI.

MILITARY ROAD EXPEDITION,

Camp at Fort Walla-Walla, W. T., October 12, 1860.

SIR: I have the honor to report the arrival of my expedition at this point on the 8th instant, and at the same time lay before the department a general summary of my operations since leaving Fort Benton.

My report to the department of August 3 gave an account of my work to that date. Having completed all the details of my reorganization at Fort Benton, and having turned over to Major Blake all the transportation that was available for his purposes, and having seen that the party to descend the Missouri were properly provided for their trip, I left Fort Benton, on my return to Fort Walla-Walla, on the 5th day of August, to carry out, as far as practicable, the instructions of the department of June 3.

As I informed the department from Fort Benton, the permanent location of the wagon road between Fort Benton and Sun river would be *via* the Teton river, on account of the advantages of wood and water; but the increase of distance can be saved during the summer and autumn months by taking the route of the Missouri river and lake. As Major Blake decided to take the latter route, I put my train in motion over it in his advance, with a view of gaining time to repair the road, and do such other work as was needed for his movement.

The coulée route referred to in my former report can be improved by work, and will yet make a good road, by heading certain coulées and ravines and working a new descent to the Missouri river, throwing the camping ground about three miles above the present point, known as the Big Coulée. We came by the coulée route, and on the morning of the 5th of August moved over a very excellent prairie road to the lake, distant thirty-five miles from Fort Benton. Fuel has to be carried to this point; water is good and sufficiently abundant for all purposes. During the spring and winter seasons the ground around the lake is wet and boggy. On the 7th we reached Sun river; no work required, and at night encamped on its right bank, at the Blackfoot agency farm. Distance from Fort Benton, fifty-nine miles.

Here we tarried on the 8th to fit up trains, &c., and moved forward on the morning of the 9th, camping at Bird-tail Rock, distance seventeen miles. The repairs of grading dry gulches, throwing out loose stones, &c., occupied the working parties during the march.

It had been represented to me that a better route to the Sun river might be found by taking a direct line from Bird-tail Rock to the Blackfoot Mission, crossing Sun river at the latter point. On this day I examined this line by going up Sun river ten miles, to the Mission, and thence over the line to the Bird-tail Rock. I found the ground rough and broken, involving many and severe changes of grade, and the line not to compare with the one selected as the route of location.

On the morning of the 10th August, moved sixteen miles, to the Dearborn river. Total, ninety-two from Fort Benton. The work consisted in grading dry gulches and ravines. The ground is generally rolling, with few changes of grade, and road good. No wood between Bird-tail Rock and Dearborn river. There is an abundance of water on Beaver creek, seven miles east of Dearborn river. This is a running stream at high water, and has a gravelly bottom ; water in pools at this season of the year. The country here becomes rolling and broken in all directions. The Dearborn river itself flows through a deep gulch in the prairie, and below, where we crossed it, enters a rocky cañon, and in eight or ten miles more empties into the Missouri. Cottonwood timber is abundant on the Dearborn, and the soil along its bottom is good.

August 11.—Left the Dearborn, and gained, in three-fourths of a mile, the high rolling prairie ground. The work consisted in grading dry ravines and gulches. Road generally good, involving changes of grade, but none so difficult as to require doubling teams. Camped on the Small Prickly Pear creek, distant one hundred and eight miles from Fort Benton.

August 12.—Moved up valley of Small Prickly Pear creek, making eighteen crossings of the same. These might be avoided by work in excavation, partly in rock, for a distance of three miles. Some of the crossings can be bridged; pine and cottonwood at hand. It will, however, be cheaper another season to work the road so as to avoid the creek crossings. It is difficult to say how much time this will involve—probably not more than five days' work of forty men; for though the rock is decomposed on the surface, it is hard and solid in the interior, and might require the slow and tedious process of blasting. It would be best, at any cost, to work the high water line.

On the night of the 12th encamped at the foot of the Medicine Rock mountain, where it was necessary to remain on the 13th to do some additional work over the mountain.

I was so unwell at this point, being obliged to be carried in a wagon, that the supervision of the work of repair devolved upon Mr. W. W. Johnson. The work here consisted in improving the grade up the mountain, which is steep, and must be changed another season, even at great cost. After gaining the steep ascent, the road is good until the descent to the Small Prickly Pear creek is made, when the road is again steep, in order to make the final descent of the most western spur of the mountain. Much work in grading was here done, both on the 13th and on the 14th, when we made the passage of the mountain and encamped at its western base on the Small Prickly Pear creek.

Morning of 15th August moved one mile and a half up a low bottom, and, with a good road over rolling ground, in two miles more reached Soft Bed creek; in two more crossed Willow creek, and at the end of twelve miles reached Silver creek, where we encamped. The bed of the creek being miry, we constructed a bridge of thirty-five feet long and fifteen wide. Graded the banks of ravines and dry gulches during the march. On the 16th August moved over rolling prairie, crossed Fir creek in four miles, and thence to the Large Prickly Pear creek, which we reached in two and a half miles more. Here we made a new crossing, and after following up its valley bottom for a mile, recrossed, and taking rolling side hills, along which the road is cut for two miles in light earth excavation, followed up the creek to its head, and in a mile and a half more reached the summit of the divide of the Rocky mountains. Descended and encamped at its western base, two miles from the summit. Placed mile-post, 149 to Benton, 469 to Walla-Walla, on the summit. At another season it would be well to avoid both crossings of the Prickley Pear creek by side-hill excavation, which will occupy twenty-five men two or three days. The ascent and descent of the Rocky mountain divide are good, and do not require double teaming.

We performed some additional work along the headwaters of the Large Prickly Pear creek and Little Blackfoot, and avoided some six crossings of the latter. The road was stony in places, and very sidling. To place this portion of the road in good order occupied the whole of the 17th, and on the 18th of August we moved down the valley of the north fork of the Little Blackfoot. Threw a bridge thirty feet long and fifteen feet wide over a slough of the stream, and then continued down its valley, grading a few points which were neglected when we first passed over it. In six and three-quarters miles, reached the Little Blackfoot proper. Crossed this stream three times in the next four miles, the road being over level prairie bottoms. We then left the Little Blackfoot, and travelling over easily rolling prairie, camped on Livingston's creek. Made 18.9 miles.

Sunday, August 19.—Left Livingston's creek, and in a half mile gained high rolling prairie ground. Passed a good camp-ground two miles west of Livingston's creek, and in a mile and a half more reached the summit of the rolling prairie and beginning of descent to the Deer Lodge valley. Descended fifteen hundred feet in the next four miles, when we reached the prairie bottom. Crossed the Deer Lodge creek four miles above the mouth of the Little Blackfoot, and camped. Light grading, where the ground was sidling, occupied the working party during the march. This crossing is ninety feet wide, with good banks and gravel bottom. Cottonwood timber lines the banks of the stream, there being no pine within five miles. The land of this valley is excellent; wheat and vegetables grow well. A small settlement is already made, and we can look forward to a dense population in this valley at an early day.

Having sent a party in advance on the 18th to explore a new line to Benetzi's creek, to avoid the crossings of the Little Blackfoot and the Hell-Gate rivers, I determined to take this line. Leaving the Deer Lodge on the 20th, we moved up the level valley of a small creek for 3.3 miles, when we ascended over rolling hills, crossed a ridge about 400 feet high, and in 3.7 miles further reached Rock creek; the only work being in grading a few gullies which were too steep for the passage of wagons. The valley and bottom of this stream are strewn with boulders, the worst of which were removed for the road. It would be well to bridge this creek, which is 30 feet wide; good rock for abutments is at hand, and timber sufficient for purposes of construction. After leaving this creek we proceeded over rolling prairie, crossing a ridge about 250 feet in height, and in a mile and a half crossed the valley of a small creek, which affords a beautiful camp-ground and an excellent place for stock. Crossed this creek, and still following easy coulées and gently-rolling ground, with but little or no work reached Benetzi's creek, at our old camp-ground of 12th July. The soil is of good quality throughout, and many fine farm sites exist all along the valleys of the streams, while the hills afford excellent ranges for stock. On Benetzi's creek building timber is found on the hills about three miles distant, and fuel on the borders of the stream. This line, *via* Rock creek, to the Deer's Lodge, involves more changes of grade than that *via* the Little Blackfoot, but it is best for the permanent location of the road, avoiding, as it does, the frequent crossings of the Little Blackfoot and Hell-Gate rivers. Distance to-day, 15½ miles.

Left Benetzi's creek on the 21st of August, and in a half mile reached the Hell-Gate valley proper; we passed rapidly down the valley over level ground; crossed Flint creek and camped on a small tributary of the Hell-Gate after having made 22 miles. The work of repairs on this portion was very light.

August 22.—Continued down the valley of the Hell-Gate, stopping for a few minutes at a small creek near the 400th mile-post to fix the crossings; hence the road is good to what is termed the "four crossing spur," where much work was done while we were here in July last. Here the road is good and permanent. At the west base of this the first crossing of the Hell-Gate is made, which cannot be avoided. The ford, which is now good, is not practicable before the beginning of July. The banks are four feet high, and firm for bridging. The structure will be 100 feet long, with a small bridge of 30 feet across a slough. Cribs will be required; good timber is at hand. The second crossing is a quarter of a mile below, and required a few minutes' work to improve it. I examined the side hills with a view of making a high water road, avoiding this and the next crossing. It will require 5,000 feet of heavy grading, part in rock, and 130 feet of corduroying to avoid marshy ground. This line, in passing westward, would, for 50 yards, then take side hills for 500 feet to gain high ground; then follows an Indian trail over sidling ground to the head of a ravine till making the final descent into the valley. The total length of grading will be 4,000 feet, transverse slope 35°, and probably will require 30 feet of corduroying, the material for which is at hand.

The line then would follow down the level bottom to sixth and seventh crossings. The fords here are good, and can be avoided by 5,000 feet of heavy grading, part in rock. It is a question whether to take the side-hill grading or bridge these two crossings, there being a point where the line may not be permanent after the grading is done. This will require especial examination.

Made eighteen miles, and camped at the seventh crossing of the Hell-Gate.

Moved forward on the 23d, and in three-fourths of a mile reached the Beaver Dam Butte. Here we worked an hour widening the portion of the road already graded; thence the line passes for 3½ miles over a level flat to the eighth crossing with slight grading at points. To avoid the eighth and ninth crossings will involve heavy work; the trail follows for two miles over steep side hills, and, in my judgment, will be practicable only at excessive cost. At many points it passes over bed rock, and for long stretches continues over very difficult ground. I examined it in person, and also the crossings of the Hell-Gate. Eighth crossing would require a bridge of 100 feet over the main stream, and a short bridge of 40 feet over a slough, with probably some slight corduroying over a bottom now dry but wet in the spring. From the eighth crossing the road follows good ground till it reaches the south fork of the Hell-Gate; here it passes over water-worn stones, and crosses the south fork in two branches. Above the forks a span of 75 feet can be had with good banks, rock for ballasting and timber convenient. The road thence for a quarter of a mile passes over good ground till reaching the next crossing. The ford is at a gravelly island where the stream is divided into two branches of 130 feet each. A bridge here of 200 feet in length will be required with heavy crib-work. Good banks can be had, and all material is convenient. The road then reaches mile-post 378, and for eight miles continues mostly through open pine timber, giving a good, firm, broad road. It crosses two small creeks with gravelly bottoms, when it reaches the tenth crossing. Here graded 40 feet from a higher to a lower plateau. Both the tenth and eleventh crossings are bad during the freshet. The high-water trail continues for a quarter of a mile on the level flat, and then takes the side hills, which it follows for two and a quarter miles over very difficult ground, following the curves and re-enterings of the spurs, which in many places are rocky. It will be best to do this grading. The bridge crossings are decidedly objectionable. The water is deep and swift, and a constant changing of channel takes place. These are the last crossings of the Hell-Gate. Thence for five miles the road is good to the Big Blackfoot. This stream, although fordable at low water, is difficult during the freshet. A bridge site can be had 150 feet above the crossing. The bridge will be 125 feet long, and will require heavy crib-work, well ballasted, to withstand the swift current in spring. The banks are very favorable and all material convenient.

From this crossing to Camp Humphreys, 25 miles, the road is good, being over a level prairie with only a few miles of gently-rolling ground intervening.

Remained at Camp Humphreys till the morning of the 27th, reorganizing the train, &c., &c., when we moved down the valley of the Bitter Root river to the Ska-o-tay creek. Here we changed the location of the road to avoid the detour made in passing up this spring. Graded steep side hill, which occupied a party of 50 men two-thirds of the day, saving a distance of a mile and a half. Next season a bridge of 40 feet in length, with 30 feet of corduroying on one side and forty feet on the other, should be built so as to make this the permanent line.

On the 28th the camp was sent forward to the "Rocky Point" region; completed the grading, &c., of the Ska-o-tay, and then proceeded on to camp. The bridge across the slough, built this spring, was in good order, and the road dry. The road through the "Rocky Points," at the eastern side of which we encamped, is very difficult, and will require much additional work to be declared in good order. There is one steep ascent of 150 feet, which it might be advantageous to avo d by building a bridge 200 feet in length, with heavy trestle-work from 45 to 50 feet in height. Fifteen cubic yards of blasting will be required, and the structure will have to be well secured to be permanent.

August 29.—Moved the train through the "Rocky Points" and made the passage of the "Big Mountain," encamping at its western base. On the next day moved to the forks of the Nemote creek; here two bridges—one 35 feet and the other 40 feet in length—were constructed. The soil of the valley is excellent; pine and cottonwood timber are abundant, and the climate in winter is mild. This would be an excellent point for farms on the main line of travel.

August 31.—Moved over "Brown's Cut-off;" repaired and improved the road, and in two miles reached the Bitter Root river. A month's work of 50 men would here save at least two miles distance, by keeping the line along the river. This involved too much work for our party to commence it this season. At present double teaming is necessary in crossing "Brown's Cut-off." At the commencement of Brown's Cut-off we found the rock that had been left this spring blasted out and the road improved, but still steep. The line might, with advantage, be thrown above, to avoid heavy grades. Road thence for a mile is good; here we reach a small creek. Three days' additional work of 30 men will be required to make the road good. Hence for a mile road passes through open pine timber to another creek, where a half day's work of 30 men will materially improve and shorten the line. Camped at Brown's prairie, 322 miles to Fort Benton, 296 to Fort Walla-Walla.

September 1.—Moved in a heavy rain-storm down the Bitter Root, work consisting in throwing out rock, making slight changes of location, and light grading at descents from plateaus, at ravines, &c. Grades are good until reaching the "Ten-mile Point," where the road will be improved by heavy side-hill grading. Made camp at the Seven-mile prairie. From this point to Hell-Gate, with the changes referred to, occupying the work of 40 men for six weeks, the road will at all times be in a good condition, needing only the slightest repairs incident upon all roads.

September 2, (*Sunday.*)—Having finished the grading of a rocky spur, moved down the Bitter Root, and in a mile and a half changed the location of a steep plateau, carried the line over a saddle; cut out considerable timber, and did some 500 yards of light grading. The line had been carried over the rolling hills, giving quite a rugged and difficult line, but by passing along the foot slopes of the ridges quite a good road was had. Did not reach camp until 4 p. m. Party moved forward, crossed the ferry, and camped on the left bank. The mile-post here is marked 273½. The descents from the plateaus to the water's edge were graded and improved. The river is now fordable one-fourth of a mile below the ferry; but as the ford is over a gravel ridge extending across the river, it may be difficult for teams to cross it. It would always be best to ferry the stream. The rope is not yet stretched across, but a good site for the same can be had. From the Bitter Root crossing to Hell-Gate, throughout the line, there is an abundance of wood, water, and grass, and good camp-grounds.

September 3.—Left the Bitter Root ferry, and, with 16 axemen and 14 with picks and shovels, began work in the timber of the Bitter Root mountains. For 3½ miles work consisted in trimming stumps and throwing out loose rocks; at the end of this distance found the road muddy and bad, requiring a corduroy bridge of 30 feet, which was built. From this point for two miles the road is good, whence to the foot of the mountain towards Cantonment Jordan the stumps require trimming, which will take a day's work of 30 men. Road over the mountain requires double teaming. It was not worked as located and marked out originally, and needs changing to improve grades. This, with work in other places, will occupy 30 men a week. It will be seen from the foregoing that the work was pushed on westward of the Bitter Root crossing, at which point the instructions of the department contemplated the construction of a bridge. This, as I informed the department in August last, is a work involving more time and labor than could possibly be put upon it at this season; and my own judgment is that, even when constructed, the advantages of it will not counterbalance the contingencies and accidents to which it must be subjected from freshets, fires, and Indian depredations. The span is 440 feet; the mean depth of water at the bridge site is 5 feet; the rise of the river is 11 feet, and no settlers within 60 miles. It would be safe to say that the cost will not be far from $20,000, or one-fifth of our last appropriation. A ferry has already been established at this crossing, which acted admirably for the passage of both Major Blake's train and our own. I have a large cable to stretch across the river during the next season, which will render the crossing easy and rapid, and which, under the peculiar condition of things in the country, will meet all the wants of travel. This will enable the means that might be expended upon this bridge to be devoted to other sections of the line where work is much more imperative.

Still continuing the work through the Bitter Root mountains, we crossed a stream 40 feet wide on the morning of the 4th September, which will require a bridge another season. Thence the road for two miles is good; following a plateau, slight repairs, such as trimming stumps, removing fallen timber, &c., were done. At the end of this distance the road rises to a high plateau, at which point the road is miry at high water. This will, in a measure, be avoided by hugging the spurs closely and grading into the side hills. Hence to the cantonment the line is good, with slight repairs of knolls and trimming stumps.

From the cantonment to the first crossing of the St. Regis Borgia the line follows good, firm ground, passing over easy plateaus, through small timber. Upon a detailed examination, found that the first and second crossings could be avoided by throwing the line nearer the foot slopes of the mountains on the left bank, following high plateaus, bridging two streams 30 feet each, and cutting through timber for one mile, which brings us to the third and fourth crossings, which are very good fords at low water, but impassable in high water. These two crossings will be avoided by a week's work of 30 men in side-hill excavation, partly in rock. Road thence to next, or fifth and sixth, crossings, distance three miles, is good; slight repairs were needed in the way of trimming stumps, removing fallen trees, and grading at a few points, the line being over plateaus covered with small open timber. To avoid the fifth and sixth crossings will require severe work in solid limestone rock for 400 feet, or bridges of 100 feet each. The excavation would be preferable. From this point to the seventh crossing (two miles) the ground is generally good, with the exception of two or three small pieces of muddy ground, which will require corduroying for, say, 50 feet each.

This brings us to the 253d mile-post. The six crossings referred to are the widest and most difficult on the St. Regis Borgia river, and by all means should be avoided by side excavation in preference to bridging. Through this section the grass is good on the small prairies and in the open timber, particularly at the first, second, fifth, and seventh crossings, and also on the side slopes of the mountains. The soil is gravelly.

September 5.—Continued the march and worked the banks of the stream at the crossings, removed fallen timber, &c. The seventh and eighth crossings will be avoided by 800 feet of heavy side-hill grading, slope 40°; perhaps some of this may be in rock. From the eighth to the ninth crossing is 300 yards, and 400 yards more to the tenth. The ninth and tenth crossings may be avoided by 600 feet of side-hill grading of the same character as previously noted. From the tenth crossing the road is good for 150 yards, when it rises to a plateau which it follows for 400 yards, when it descends to the eleventh crossing with grading, which must be bridged. The span is 110 feet, and the banks are good. Continued for one-third of a mile, and made the twelfth crossing. The bridge here would be 100 feet long; the left bank is sufficiently good; the right bank will require an abutment. The general rise of the water at all these crossings is between three and four feet. Line hence for 80 yards passes through a willow thicket which needs to be corduroyed. In a few hundred yards made thirteenth crossing, which will require a bridge of 100 feet; the banks are good. Thence for a half mile continued on right bank; road good. Did much work of clearing, trimming stumps, and grading. Made only two and a half miles, and did not quite complete the road up to camp; built two small bridges and had the road completed up to this point, the 250½ mile-post, on the morning of the 6th.

This section of the road last year was worked under the marked disadvantages of snow and frozen ground. The fourteenth crossing needs a bridge of 150 feet, in two spans of 65 feet each. The line between the crossings is generally good, but a few slight changes of location may improve it. Thus, though frequent crossings of the stream are now made, only four bridges will be required. When this and the side-hill work is done, 20 miles of the road west of the Bitter Root river will be in excellent condition.

September 6.—Moved forward, and in three-fourths of a mile made the seventeenth crossing, which is 120 feet wide; the banks are good; bridge required. Road thence in 300 yards makes the eighteenth crossing, which will require a bridge of 110 feet. Road follows left bank for 300 yards and makes the nineteenth crossing; banks are good; a bridge here of 110 feet cannot be avoided. In 150 yards, road reaches a large branch, which it crosses just above its mouth, and then makes the twentieth crossing, which will require a bridge of 95 feet span. Graded side hills for 100 feet, and in a quarter of a mile made the twenty-second crossing. The ground here needs to be examined in much detail to determine the amount of excavation required to avoid the upper crossings; twelve crossings may be avoided by heavy side-hill grading and 24 bridges.

Encamped at the thirty-sixth crossing on the right bank, and marked mile-post 244. This point is what is known as the Five-mile prairie.

Moved on the morning of the 7th, and in 600 feet made the thirty-seventh crossing; and in a half-mile more reached the thirty-eighth, which was altered; the road is good. From here to the next crossing is a half mile; good bridge sites at both points; spans 80 feet each. Continued on the left bank for a half mile, and reached a difficult point, where we were obliged to build a corduroy bridge of 60 feet over marshy ground; and reached, in a quarter of a mile further, several crossings where the stream crosses and recrosses the valley, near the foot of the mountain. These are all difficult, and will be avoided another season by either heavy grading, part in rock, or by a long heavy bridge, in order to gain a high plateau. The road next makes the second crossing from the foot of the mountain, after which it continues for three miles over high plateaus. Another season's grading will be needed, and the stumps will have to be re-trimmed. Reached eastern base of the Cœur d'Alène mountains, and made the passage with one upset; road generally good.

On the western descent of the mountain some springs have broken out into the road, rendering it miry; the location will be improved by throwing the line some 50 feet to the north, into the side hill, making a good, dry road. We found but few land slips on the road where it was graded along the mountain side last season. It will need repairing another season. The plateau at the foot of the divide, and the road generally on this side of the mountain, need working. The second curve of the road along the mountain side needs changing, and a general overhauling of the whole line should be made before the line can be considered as completed for travel.

We worked twelve continuous hours without even a lunch, and camped at night on the second small creek from the foot of the divide.

September 8.—Started at an early hour, and in 300 yards reached a large creek coming from the north; changed the crossing here, and in a quarter of a mile more reached a point where the river had broken through its banks and taken the road-bed for its channel. Fifty yards of heavy side-hill grading will avoid this point—say, two days' work of 50 men. The line, for a quarter of a mile beyond this point, is sidling, and will require grading another season; thence to Long Prairie the road is good; stumps will require trimming. The question had been with us, at one time, whether to take the road *via* "Johnson's Cut-off," so as to avoid all the upper crossings of the Cœur d'Alène, or bridge these crossings. I made especial examination of both routes, and found the bridge work, if made strong and secure, would be by far the best, giving an easy and nearly level road.

Road for a half mile from Long Prairie passes over good ground, through dense, but not large timber, to the first crossing, which will require a bridge of 25 feet span. The banks are good; timber at hand. In 100 yards made next crossing, which will need a bridge of 30 feet span; the banks are low. In 100 yards made third crossing; bridge required of 30 feet; banks are 4 feet and 2½ feet high, respectively. In 200 yards reached fourth crossing; bridge 35 feet span; timber here is not very convenient; good rock for ballasting. 100 yards to the fifth crossing; bridge required of 30 feet; banks are good; rock at hand. In 100 yards made sixth crossing; 35 feet span; good timber, and rock convenient. The seventh crossing was bridged last season; the structure is in good order. In 200 yards made eighth crossing; bridge of 25 feet required, and in 100 yards more made the ninth. These two might be avoided by side-hill work, if not, will require a bridge of 40 feet across the latter; banks are good. In 200 yards, through cedar timber, made tenth crossing; bridge of 35 feet required; banks are good. 100 yards further made eleventh crossing; bridge needed of 30 feet span. 80 yards hence to twelfth crossing, which will need a bridge of 30 feet; banks good; rock convenient for ballasting. This brings us to Johnson's Cut-off, where we encamped. Marked mile-post 229. The road between the crossings above noted is good, and needs but the trimming of stumps and repairs at mud holes to be at all times good. Excellent grazing is to be had on the side hills and up the ravine at this point, which expands into a basin some 300 yards in diameter and half a mile in length, enclosed by mountains from 1,500 to 3,000 feet in height. This would be a good point for trains to encamp, and recruit stock if necessary. Animals can also be driven by the trail over the hills to this place from Long Prairie, which is only a mile distant from the head of "Johnson's Cut-off."

Morning of the 9th, repaired a bridge over a slough at the 13th crossing, 80 feet in length. This is the quicksand crossing referred to in previous reports. This crossing will require a bridge of 40 feet in length; the banks are not high. The road from Johnson's Cut-off to the forks of the Cœur d'Alène river (5 miles) involves fourteen crossings of the stream. The work of another season in this distance will be repairs of corduroying and mud holes, in addition to the bridging, which in detail are as follows: The fourteenth and fifteenth crossings are bad, requiring bridges of 30 feet each, with 30 feet of additional corduroying. The sixteenth crossing will require a bridge of 50 feet span diagonally across the stream; timber which has been killed by fire is at hand for the purposes of construction; rock also for ballasting. Seventeenth crossing, bridge 25 feet long; banks good. Eighteenth crossing, bridge 30 feet. Nineteenth crossing, bridge 35 feet. Twentieth crossing, a clump of large cedars on the right bank; span 30 feet; cedar timber for structure. Here a change of location from a lower to a higher plateau will be needed. The twenty-first and twenty-second crossings might be avoided by heavy excavation 200 yards in length; but as nearly all the side hills have a slope of from 30° to 40° and show outcropping rock, it will not be found always practicable to take the line by them; hence the absolute necessity of bridging all the crossings, with the few exceptions noted. The twenty-third crossing has a span of 30 feet, and needs corduroying on both banks. The twenty-fourth crossing needs a bridge of 30 feet. Here we enter large cedars, where the bottoms are wet and will need repairing at all times. Twenty-fifth crossing has a span of 30 feet. Twenty-sixth crossing, bridge of 30 feet Here examination is required to avoid a cedar swamp and muddy ground. Twenty-seventh crossing has low banks which can be spanned by a 30-feet bridge; timber is not convenient, unless cedar is used. Twenty-eighth crossing, 30-feet bridge; banks good; timber convenient. From the forks of the river at the 224th mile-post to Mud Prairie, 15 miles, the work of another season will consist in bridging streams, removing such timber as may fall during the winter and spring, and repairing muddy places. We moved, September 10, to within four miles of Mud Prairie, doing much work in repairing and grading the banks at the crossings. The road was, however, in generally good order, except at a point of side hill within one mile of Mud Prairie, to which latter point we moved on the morning of the 11th. As the road here was considerably out of repair I deemed it best to remain over in camp at the Mud Prairie on the 12th to repair it.

Taking up the line at Mud Prairie, and going eastward to mile-post 224, at the forks of the Cœur d'Alène river, the line and bridging are as follows: In one mile reach the second crossing of the river above the Mission, which will require an 80-feet bridge; from 100 to 300 feet of corduroying may also be necessary. Thence, for three miles, pass over good ground, through timber of medium growth, and reach the third crossing, where work to be hereafter referred to was done, and which involves a bridge of 90 feet in length. In 200 yards more reach the fourth crossing, which will require a bridge of 50 feet span; road needs repairing, holes to be filled up; changed the site of the abutments on the left bank; rise of water here is about three feet; left bank here is three feet above water level, right bank two and a half; ballasting rock is not convenient, unless gravel is used from a bar close by. In a quarter of a mile the fifth crossing is reached; span 50 feet; right bank is four feet high, left bank from two and a half to three and a half feet. The road on the left bank will have to be changed somewhat to suit the bridge crossing; ground is low and flat and corduroying may be necessary. Sixth crossing has good banks; span is 45 feet; gravel is convenient for ballasting. Seventh crossing has low banks; 50 feet span required; cedar timber is abundant, of large size. Eighth crossing will need a bridge of 80 feet; a centre crib will be necessary; the left bank is low and marshy for 30 feet, the right bank is high. This crossing should be avoided, if possible. I examined the ground in person, but found that the line would pass through cedar swamps and is almost impracticable. Ninth crossing will involve a bridge of 50 feet; right bank is four feet and left bank two and a half feet above water level; cedar and pine for bridge. The ground was also examined with a view of avoiding the 10th crossing, but found the work impracticable, except at great cost. A bridge of 50 feet will be required here; the right bank is four feet high, left bank two feet. Eleventh crossing will require a bridge of 50 feet, with 50 feet of corduroying on the left bank; a small creek, before reaching the 10th crossing, will have to be bridged also. In half a mile more a branch from the south is reached; span is 90 feet, with intermediate supports required. Twelfth crossing will require a span of 50 feet; left bank four feet, right bank two and a half feet high; rise of water three feet. Between the south branch and twelfth crossings 100 feet of corduroying will be necessary. Thirteenth crossing has two branches, which require bridges of 50 feet each; main banks three feet high, intermediate banks one and a half feet high. From the thirteenth to the fourteenth crossing, a distance of four miles, the road is excellent, passing through small prairie openings and medium timber; but few repairs will ever be required at this point. Fourteenth crossing will require a bridge of 50 feet; left bank three and a half feet high, and right bank two and a half feet high; rise of water two feet; good timber on right bank. In one mile more the fifteenth crossing is made, 40 feet wide; banks good. Sixteenth crossing will require a bridge of 40 feet span; right bank is five feet high, left bank two and a half feet high. The road between the fifteenth and sixteenth crossings will need corduroying in places another season. Here we enter another section of cedar bottoms, which is wet and miry in the spring and summer. Much additional work will be here needed to make the road what it should be. From this point onward I examined the side hills on the right bank with the view of avoiding the first crossing of the North Fork and the several small branches into which the stream divides itself about the point of junction with the South Fork, and which are bad and difficult during the spring. I found the spurs rocky and precipitous indeed, rendering the exploration so difficult that I had to wade down the bed of the stream in order to get along at all. It is out of the question to take these side hills as the line of location, except at greater cost than we would feel warranted in expending upon this section of the line. The crossings, both of the main branch and the smaller ones into which it divides itself, will have to be bridged and the ground corduroyed wherever required. This done, the line will be good; at present it forms one of the most difficult sections of the line. The bridges across the three small branches will be 25, 30, and 40 feet each, respectively. The South Fork will require a bridge of 30 feet, at mile post 224–394; a half mile above is the first crossing of the North Fork, where a bridge of 70 feet will be required. The banks at these streams are low; timber, as usual, is abundant; no rock will be required for ballasting; rise of water is about two feet.

Thus it will be seen, therefore, that the principal work both along the St. Regis Borgia and Cœur d'Alène rivers will be in bridging the streams, and the natural repairs incident upon the first opening of any road. This, indeed, forms the principal difficulty upon the whole line. Special study was given to this line along these rivers, and the foregoing forms about as approximate an estimate of the true amount of work as we could gather. As this will be again referred to in a report of estimates of labor, &c., to be done next season, I will pass on and resume the itinerary of operations on our route from Fort Benton.

However, before leaving this subject, I would remark that it is exceedingly difficult to determine the rise of water in these two streams, except at three points upon the Cœur d'Alène river, one mile below "Johnson's Cut-off," where the highest high-water mark is four feet above the present or low-wa er stage; the second some five miles above "Mud Prairie," where the highest rise was about three and a quarter feet; and the third at the crossing above the Mission, where the rise is seven feet. The banks do not appear to be ever overflowed except where only a foot or so above the water, and then for only a few yards back from the stream. On the St. Regis Borgia the water has a rise of about three feet, but no positive reliable flood mark was seen.

Having completed the repairs of the road at the foot of the spur above Mud Prairie, before referred to, we moved forward to the first crossings of the Cœur d'Alène river. The work consisted in building a bridge across a slough in Mud prairie, 25 and 15 feet, corduroying 50 feet near Seven-mile Prairie Saddle, and did 50 feet of side hill excavation, brushed the road from point to point where the ground was muddy, besides other light repairs. Camped on the 13th at the first crossing of the river on the left bank. Grass on this section is excellent, and the road, with a few repairs at another season, will be good. It had been my intention on reaching this crossing to select a bridge site, and should the work not prove too severe, endeavor to build the pier and abutments this season. This crossing is the deepest and the last to be fordable of all the crossings, and either a ferry or bridge is needed for travel earlier than July. The best site for the bridge would be about one hundred yards below the present ford. The banks are 10 feet high, and the width of the stream is 148 feet. The water is five feet deep, and rises seven feet. By extending the structure some 40 feet each side of the river high ground will be secured. Timber is near at hand. Without the aid of blocks and strong tackle I found it not feasible to begin such a heavy structure as this would involve; and as, in my own mind, the time at our disposal was to be consumed either in bridging this crossing or the third, and finding here the water not swift, so that our flatboat could be used to advantage, I deemed it better to construct a heavy pier and abutments at the third crossing. Having determined upon this, I sent Lieutenant Lyon forward with the larger portion of the expedition to repair the road onward to Walla-Walla, and with twenty-five men remained behind to complete the work. We were occupied ten days at this point, or till the 24th of September. The width of the stream was 96 feet. We constructed a triangular pier 24 by 21 by 13 feet, and seven feet above the present stage of water. The abutments were 18 feet wide, and about six feet high above water level. The depth of the stream was 18 inches, and its rise about three feet. The spans are 40 feet each. Both the pier and abutments were filled with rock quarried from the mountain sides near by, and will remain secure either against freshet or drift. Having completed this much of the bridge, the stringers not being thrown over for want of blocks and tackle, we resumed our march towards Fort Walla-Walla, encamping on the 24th at the Mission. Having placed in store at this point such property as was not needed for our present purposes, but which would be available another season, we continued our march along the right bank of the Cœur d'Alène for nine miles to the ferry. Road was good. Crossed and moved down left bank to the mouth of the ravine leading to the St. Joseph's. Road good, grass excellent. The advantages of this valley, as referred to in previous reports, cannot be over estimated, and are such as will attract at no distant day a hardy population to its cultivation.

On the 25th moved over to the St. Joseph's river, crossed and encamped on the left bank. Road also good, excepting the steep places, which will be improved by work. The grazing and soil in this valley are excellent.

On the 26th moved on, crossed the St. Joseph's bridge, which was left in good repair by Lieutenant Lyon, and after travelling eighteen miles encamped at the springs west of the Lochultz. The road was good; grass was, however, burnt; timber at hand for fuel.

Special examination was given the water marks along the St. Joseph's. Near the bridge the signs of water are such that the rise of water must be not less than eight feet, or from two to three feet above the banks, indicating that the entire lower portion of the valley becomes a lake at high water. I am convinced that it will be impracticable to make a high water road by following the present location *via* the southern shore of the Cœur d'Alène lake and valleys of the St. Joseph and Cœur d'Alène. We passed over this route in July without difficulty; but all things since go to show that a change of location must be made in order to have a road capable of being travelled at all seasons. The present route will always be practicable during low water, and is admirably adapted to all the wants of travel, and as an emigrant route, supplying the wants of grass, water, and fuel, will not be surpassed. During the spring and freshet seasons the line from the high prairie region to the west of the lake to the Mission cannot be travelled. This section cannot be changed and thrown to the north of the Cœur d'Alène lake. This was feared last year, but then it was not intended to make the road by this route, but in obedience to instructions from the department the line was carried as located.

With the view of ascertaining the amount of work required for the high water road, I despatched Mr. Johnson, on the 14th September, with an Indian guide, to explore the line from the mission to the outlet of the Cœur d'Alène lake. His report, which is being prepared, and which will be forwarded by the next mail, would show that no less than two months' work of fifty men will be required on this section of twenty-eight miles, to make it practicable for wagon road purposes. The line passes for half this distance through a densely timbered region, and for the remainder through open pines. The changes of grade are frequent, and in some places severe, except at great cost. The general line is that travelled by Colonel Wright, in 1858, and which is given in the memoir submitted to the department in the winter of that year. *This will have to be the line of the road.* From the outlet of the Cœur d'Alène lake the road passes through an open timbered level country to a good crossing of the Spokane river at Antoine Plantes. From this point Mr. Johnson followed what is known as Mix's road, which, passing over good ground, comes into our road thirty-five miles southwest of the Spokane at the Sil-sil-pow-vet-sin. Thence onward, it follows good level and dry ground, passing between the two lines travelled by Colonel Wright, in 1858, on his passage into and from the Spokane country. This line avoids all the crossings of the Pelouse and its tributaries, skirts the Pelouse at its great bend, fourteen miles from Snake river, and strikes the Snake river at the mouth of the Pelouse. On it there is an abundance of water in the spring, is well grassed throughout, and, in connexion with our present road, to be travelled during low water, is sufficiently supplied with fuel. It subserves, therefore, many advantages, and forms a beautiful connecting link with our line onward to the Mission and beyond. I will again refer to this in my report of estimates of labor, &c., which will be prepared as soon as I complete the present report.

September 28.—Moved westward ; road is good, following, easily, rolling prairie ground and level bottoms and flats for eleven miles, when we encamped upon the Ned-l-huad-lk creek, 138 miles from Walla-Walla. Grass was burnt, but found sufficient for our animals.

September 29.—Continued westward for 17 miles, over good and easy ground, to the Sil-sil-pow-vet-sin, where we marked 121 miles. The whole line being repaired by Lieutenant Lyon, little was left for us to do. As a detailed description of this portion of the road has been given in previous reports to the department, I simply give an itinerary of our marches.

September 30.—Moved forward to the Tcho-tcho u-seep, nine miles, or 112 from Walla-Walla, and camped. Road and camp ground both good. No water in this interval.

October 1.—Moved on to the O-ray-tay-ouse, 96 miles from Walla-Walla. Road good.

October 2.—Left the O-ray-tay-ouse, and, with a good, nearly level road, reached the Mo-cah-lis-sia in 13 miles, and in one more the Pelouse, which we crossed and encamped on its right bank, 80 miles from Walla-Walla. Grass here not burnt.

October 3.—Moved down the Pelouse, crossing it three times at good fords, and encamped on its left bank; grass and wood scarce, but road good. Our original line follows the high rolling ground to the Snake river, opposite the mouth of the Tukañon, but with the change of location to be made, as before referred to, we determined to follow the general direction of the Pelouse as far as practicable, and cross Snake river at the mouth of the former.

October 4.—Moved down the Pelouse, crossed the stream three times in five miles, reached the Colville road, and in 12 miles more reached Snake river, crossed and encamped on the left bank. The road to this point is generally good, and, with few repairs, at another season will be an excellent line.

Moved on the old road on the morning of the 5th to the Touchet, where we found the main portion of the expedition encamped, engaged in bridging the stream. As the bridge was not completed, determined to remain until all was finished except the flooring. Bridge 110 feet by 16 feet, with two cribs six feet high; four stringers with 65 feet of corduroy on right bank. This occupied us till the morning of the 8th. As there was nothing but cotton-wood at hand for a covering, concluded to haul out pine plank from Walla-Walla for the purpose. Moved in to Walla-Walla on the 8th; discharged such of the employés as were no longer needed, and began to prepare for our winter's abode.

The line from Walla-Walla to Snake river, *via* the Toukañon, is preferable on account of wood and water, and, indeed, the crossing of the Snake here is equally as good as that at the mouth of the Pelouse, but when it is remembered that, on account of the rapids and rocks which obstruct the Snake river just above the mouth of the Pelouse, steamers can ascend to the latter point much later than to the mouth of the Toukañon, and this line forming the present travelled road to Colville and northward, avoids the crossings of the Pelouse, which, being swollen in high water, are difficult, and also from the entire absence of timber fit for purposes of construction, can be bridged only at great cost, it would seem to be judicious to take the line direct to the mouth of the Pelouse. The disadvantage of a twenty-seven mile march, without water, from the Touchet, is objectionable, and to trains very serious. There is a spring on the road, seven miles from the Touchet, which, by improving, may supply water to trains. I shall attempt this either during the winter or spring, by preparing tanks, &c. The road needs to be otherwise worked and improved. From the Touchet to Walla-Walla the road is good.

The foregoing will give a general account of our work and operations since leaving Fort Benton, and points out the character of work yet to be done to make the line permanent. Detailed notes of amounts and character of future work have been made, which will form the basis of my report of estimates of work, which I will prepare and submit to the department by next mail.

I cannot conclude this report without referring to the successful passage of Major Blake's command over our road. His views will be doubtless submitted in a report to the department. For myself, I feel gratified in seeing the changes that a year has already wrought at points of the line. A settlement of seven houses when we started has grown to seventy on our return to Walla-Walla. Farms have been opened along the line, and, with the improvements that we propose putting on the road, I am sanguine to see ere many years a continuous line of settlements from the Columbia to the Missouri.

I am, sir, truly and respectfully, your obedient servant,

JOHN MULLAN,
1st Lieut. 2d Artillery,
in charge of Military Road Expedition.

Captain A. A. Humphreys,
United States Togographical Engineers,
in charge of Office of Explorations and Surveys.

XVII.

Military Road Expedition,
Camp at Fort Walla-Walla, W. T., October 25, 1860.

Sir: I have the honor to submit to the department the following estimate of time, labor, and expense yet needed to be applied to the construction of the military road from Fort Walla-Walla to Fort Benton, and at the same time a plan of operations for the next season, which, formed from all the data and facts before me, has seemed to be judicious and economical, and which I am sanguine to believe can be successfully and fully carried out.

As the department will gather from the report submitted October 12, 1860, it has been found impracticable to carry the line to the south of the Cœur d'Alène lake, so as to make a good road for travel in spring and during the freshets; and hence this section of the line must be thrown further northward around the *northern* rim of the lake. This involves then a direct connexion of the head of the Spokane and the mouth of the Pelouse.

Such being the inevitable necessity for a high-water road, the line following from Fort Walla-Walla to the mouth of the Pelouse will be the line we travelled on our return from Fort Benton, and which is described in my last report.

Crossing Snake river, it will follow the general direction of the Pelouse to Cur creek, which it crosses at its mouth, and thence avoiding all the crossings of the Pelouse; and its tributaries will pass over an easily rolling prairie and plateau region till it strikes our old road at the crossing of the Cil-sep-po-vest-sin creek, which will not require a bridge; thence on to the Nedlwhnalk, which it crosses seventeen miles below our present road-crossing; thence in thirty-three miles it reaches and crosses the Spokane, either at Antoine Plant's or where Colonel Wright crossed it in 1858; thence *via* the Spokane and northern edge of the Cœur d'Alène lake to the Mission. Mr Johnson's detailed report, submitted herewith, explains the whole line and the exact character and amount of work to be done.

I propose, therefore, to start from Fort Walla-Walla by the 1st of April, reaching the Spokane by the middle of April, from which time to the 1st of July, the earliest date at which any work of bridging along the Cœur d'Alène will be practicable, we will be engaged in making the high-water road from the Spokane to the mission, and which I am sanguine to believe will be ample time, unless some unforseen contingencies arise.

The following is the minimum force for the accomplishment of this work; and the figures stated are the figures now ruling at this point, except for mechanics, which I fear is too low, though obtained last year at that figure; and it is for this reason that I have retained as laborer and carpenter Hugh McEnen, who has been out with me for eighteen months, and who is willing to go at the same figure next season; and being an invaluable man on the road, he is with me herding, &c., till spring shall enable us to organize:

2 engineers in charge of parties, at $125 per month..	$250 00
1 physician, at $125 per month....	125 00
1 wagon master, at $100 per month....	100 00
1 clerk and commissary in charge of property, at $125 per month....	125 00
1 blacksmith, at $75 per month....	75 00
2 carpenters, at $75 per month....	150 00
12 teamsters, at $50 per month....	600 00
4 cooks, at $50 per month....	200 00
35 laborers, at $50 per month....	1,750 00
Hire of extra duty men....	500 00
Making a monthly expense of....	4,000 00

All of which service can be employed at this point at the date of organization.

The amount on hand at the end of October, after paying out all outstanding indebtedness will be, say....	$5,637 50
Accruing from sales of property, say....	9,362 50
Amount yet to be deposited of the appropriation....	70,000 00
To which, add value of rations on hand, say....	2,500 00
Total....	87,500 00
Probable expense from November 1 to April 1, pay and rations of employés, fuel, &c., say....	2,500 00
Leaving on 1st of April, available for road....	85,000 00

With this amount, I think it safe to say that, by exercising a rigid economy and adhering to the figures below given, we shall be enabled to outfit our expedition and occupy the field fifteen continuous months, or from the 1st of April to the 1st of July, at which time we shall reach and connect with the American Fur Company's steamer at Fort Benton. My party disbanded and its discharged members guaranteed a safe and pleasant return to the Atlantic settlements

The approximate estimate of time and labor would be therefore as follows:

15 months' service, at $1,000 per month	$60,000 00
3,000 rations, at 35 cents	10,500 00
6 wagons and rigging complete	2,000 00
60 yoke of oxen, at $120	7,200 00
70 yokes and cows, $500; chains, &c., $300	800 00
Tools, consisting of axes, picks, shovels, blasting, carpenters' and blacksmiths' tools, nails, spikes, and iron crowbars, &c., say	2,000 00
Camp equipage, cooking utensils, &c	500 00
Expresses, ferriage of Snake river, transportation	1,500 00
Contingent and incidental expenses	500 00
Total for fifteen months	85,000 00

In this estimate I have allowed only for those things which will be absolutely necessary next season, and though the item of contingency be small, I am sanguine to think we can work within the limits as given. I have given the subject of our next seasons' movements much study, and believe that the work can be placed in a condition satisfactory both to myself and to those who will use it, by adhering to the views herein contained.

Starting then with our expedition on the 1st of April, with eight month's supplies, we shall reach the eastern base of the Bitter Root mountains by the 1st of May, where our heaviest work begins. From this point our heavy trains can return to the mouth of the Pelouse and bring up our remaining seven month's supplies, getting back by 1st of June.

By the 1st of July our work and the supplies for the entire period will be at the Cœur d'Alène Mission. Here our bridge work begins, and from this point to the Bitter Root river we shall have not less than eighty bridges, as shown by my report of October 12, 1860, in addition to the side-hill work to avoid other crossings.

Dividing my working force into four parties, I have estimated to be able to bridge each crossing in six days, or with my entire force four bridges in six days, and allowing for bad weather and moving camps, say 16 bridges a month, or a period of five months for the 80 bridges, or until the 1st of December. This can be done, however, together with the repairs of the road along the line as now worked. This period will then bring us to the Bitter Root river, and to the commencement of winter.

During this interval, the road being already opened in our advance, we can occupy our trains in moving forward our supplies to the Hell-Gate Ronde. At the end of our Bitter Root mountain work our animals will be much reduced when they can have the genial climate and fine grasses of the Bitter Root valley for the winter.

Our main camp then, in December, can be established at the crossing of the Big Blackfoot river, which requires, as before reported, a structure requiring much time and labor. From this point I can distribute my working parties along the several crossings of the Hell-Gate, referred to in much detail in my last report, and thus continue the work of bridging, occupying January, February, March, and April, and, after allowing for the contingencies of snow and unpleasant weather, I believe this will be sufficient time for all the bridge work, giving us May and June for the side-hill excavations and other work to Fort Benton.

Thus it will be seen that every month of the period will be occupied by our parties in the field; but if we are compelled to move forward and work the road in the Bitter Root range until snow drives us backward, not one-half the amount of work can be done as we could accomplish in moving forward and wintering our stock in the Bitter Root. My own conviction, then, is to occupy the field for 15 months continuously, and reach Fort Benton 1st of July, 1862, discharge the greater portion of my party, and with the remainder and my escort return over the road for repairs, &c., to Fort Walla-Walla, and thus put in complete order the entire line opened during the seasons previously.

The supplies for my parties on reaching Fort Benton, needed for those returning and those descending the river, can be forwarded to Fort Benton in the spring of 1862, and which will reach there about the date of our arrival.

I have in the foregoing endeavored to lay before the department all the details of the work along the line; and the method which, after studying the ground for the last two years, has appeared to me to be the most judicious and economical to pursue in accomplishing it. And I have placed the estimates low and at figures now ranging in the country.

With the outfit stated, I feel warranted in guaranteeing the final success of our work.

I would desire the department therefore to early consider my views and plans so that I may be notified at the earliest dates.

It is highly important that I should be apprised of the action of the department at the earliest day practicable, in order to prepare for my outfit for an early start in the spring, as it is essential that I should leave this point by the 1st of April.

Our mail facilities to Fort Walla-Walla are much interrupted during the winter, having but one monthly mail. I would desire therefore that all my mail communications be sent to me at Fort Vancouver, Washington Territory, at which point I may meet them in person. I trust to hear from the department by January or February.

I would at the same time request that $30,000 (thirty thousand dollars) of the appropriation be placed to my credit with the assistant treasurer in New York city, and that I be informed of the same by the same mail, in order to prepare my outfit and organize my expedition in time to take the field by the 1st of April next.

I am, sir, very respectfully, your obedient servant,

JOHN MULLAN,

First Lieut. 2d artillery, in charge of Military Road Exp.

Captain A. A. HUMPHREYS,

United States Top. Engineers, in charge

Office Explorations and Surveys, War Department.

A.

WAR DEPARTMENT, *Office Explorations and Surveys,*
Washington, D. C., March 15, 1859.

SIR: An appropriation of $100,000 having been made for continuing the construction of the military road from Fort Walla-Walla to Fort Benton, the Secretary of War directs that your former plan of operations for opening that route be modified to suit the more permanent character of road contemplated by the department in recommending to Congress an appropriation for its construction.

In conducting this operation, your attention will first be directed to making those parts of the route where the greatest difficulties and most numerous obstructions exist practicable for the passage of wagons at all seasons of the year.

Immediately upon the receipt of this communication you will report the nature and extent of the outfit necessary for the purpose of building the road, and its probable cost; the number of assistants and others whom you deem it necessary to employ, with their rates of compensation; and the expense of organizing and maintaining the expedition in the field during sixteen months. Upon this plan being approved by the department, you will, with your assistants and such employés as it may be found advisable to engage here, proceed to Fort Dalles, Oregon, *via* New York, where you will have collected and prepared as soon as practicable the tools, materials, and outfit requisite for the work, and employ such mechanics, laborers, and other persons as may be necessary for the service.

You will locate the road from Fort Walla-Walla to Fort Benton, as laid down in previous instructions.

The commanding officer of the department of Oregon will be directed to detail one hundred (100) picked enlisted men of the infantry or artillery, including a number of mechanics, with three subalterns, and order them to report to you for duty either at Fort Dalles or Fort Walla-Walla, as he may find best for the interests of the public service.

The officers of the quartermaster's, subsistence, ordnance, and medical departments serving in the department of Oregon will be instructed to furnish the expedition, upon your requisition, transportation, quartermaster's stores, arms, ammunition, &c., provisions, medicines, and medical stores, the articles for the use of the civil employés being paid for by you out of the appropriation for the road.

So far as it can be done consistently with the proper protection of the work, you will cause working parties to be detailed from the enlisted men of your command to aid in the construction of the road, who, while so employed, will receive the extra pay allowed by paragraph 883, Army Regulations.

You will make such reconnaissances as may be necessary for the proper location of the road, as also the usual surveys over the routes pursued by the expedition, and, so far as it can be done without interfering with or adding to the expense of the accomplishment of the special object for which the appropriation was made, every opportunity will be availed of to gain information respecting the region over which your movements will extend.

Such expenditures for books, small instruments, and repairs of instruments as you may find necessary for your purposes you are authorized to make.

Upon the completion of the work you will return by such route as the condition of the party may render necessary or desirable, discharge your employés, dispose of your outfit at the most favorable and convenient point, and repair to Washington with such assistants as may be required to complete your office work.

Very respectfully, your obedient servant,

A. A. HUMPHREYS,
Captain Topographical Engineers, in charge.

Lieutenant JOHN MULLAN,
2d Artillery U. S. Army, Washington, D. C.

B.

Report of survey of the wagon road from Fort Walla-Walla to Fort Colville, made June, 1859.

CAMP ON PELOUSE RIVER, W. T., *July* 7, 1859.

SIR: After receiving your instructions to accompany Brevet Major P. Lugenbeel's command to Fort Colville, for the purpose of making a survey and map of the wagon road and the country *en route*, I left the Dalles on the 16th of May, 1859, and joined the command, which consisted of two companies of the 9th infantry, United States army, at Mud Springs, on the same evening, from which point we started on the 18th and travelled over the usual route to Fort Walla-Walla, meeting with no other difficulties except those occasioned by the high water, which made ferrying and fording the different streams and creeks more laborious than it would have proved at a later season. We reached Fort Walla-Walla on the 27th of May, and, after all the necessary arrangements were completed, we left on the 1st of June. Captain Kirkham, of the quartermaster's department, who was instructed to carry stores to Fort Colville, had started in advance of

Major Lugenbeel's command with 32 six-mule teams and this train, keeping all the way in front of our party, benefitted us very much, as we had only to follow their tracks, saving us the trouble and delay to look out for the best location for the wagon road, and enabling us, besides, to avoid bad portions by making little detours. The road from the Dalles to Snake river having been surveyed and reported on by you previously, according to your instructions, I commenced my work only at the north side of Snake river. Laboring under the impression that you had also, in 1858, taken the direction towards the mouth of Pelouse river, I neglected to run a compass line over the country between the Touchet and Snake river, but having travelled twice over that section of the country, I am satisfied that I can give a correct description of it. I submit to you this report in journal form, as it will give me an opportunity to go more into the detail of each day's travel.

June 1.—Left Fort Walla-Walla early in the morning, and making a detour towards the west, on account of a wet place which made the usual route almost impracticable for heavy loaded wagons, we struck Dry creek at the usual crossing place Following from there the well marked road we reached the Touchet river, and fording it without difficulty, encamped on its north shore, having made 21.4 miles. I have to mention that my odometer measurement differs from yours by 0.7 mile, originating in the westerly detour we were obliged to make.

June 2.—Started at 5 a. m., and followed, for four miles, a good level route on the north side of Reed Creek valley. There, turning short to the north, the road led through a winding and gradually ascending valley, with a dry water run. Two miles from the mouth of this valley we discovered a small spring in a little side ravine, containing little, but very cool water. The road then turning to the northeast led through a narrow and rather rocky cañon, still ascending. At some points this cañon would widen a little more and close up again at others, alternating in this manner for four miles, when a high prairie ridge, which has to be surmounted, closes it up. The ascent and descent of this ridge are not easy, but very practicable for any kind of vehicle, and the top of it, which we followed for two miles, winding about, according to its formation, offers a magnificent solid road. The valley leading towards Snake river affords also a good road, and bears in a general line about north by east, and extends for twelve miles. The final descent to Snake river is accomplished by winding around the several bluff points, and is by no means an easy one for wagons. Distance travelled 26 miles. The want of water and of shade is keenly felt on this long and tedious stretch, and renders it hard on men and animals. At Snake river we found two ferry boats, one of them belonging to the government and the other one to a private company; the latter one will be transformed into a flying ferry, and in this form will be of great use to the travelling community. The width of Snake river at the present high water is about 350 yards, and the current exceedingly rapid and strong. By a combination of adversities we succeeded in getting everything across only on the evening of the 6th, and took up our line of march.

June 7.—At $5\frac{1}{2}$ a. m. The ascent to the table land extending towards the Pelouse river is exceedingly difficult, the bluffs being from 900 to 1,100 feet high, very steep, and consisting, at those places where the rock does not outcrop from the surface, of a sandy soil. The river bluffs, which the road ascends, forms four terraces, of which the first one is about 500 feet, the second 700, the third 900, and the last 1,100 feet above the river. After reaching the fourth plateau the road gains almost level but sandy grounds, and keeping along the easterly base of a steep prairie ridge, strikes, in four and one-half miles, a ravine, which has a gently declining slope. In four miles we reached the mouth of the ravine, and saw before us the deep cañon through which the Pelouse river forces its way. The almost perpendicular bluffs, which hem the river forbid any attempts to gain the valley of the Pelouse, and the pedrigal character of the plateau forces the road to make a detour of two and a half miles before the upper river bottom can be gained. I learned by information that the Pelouse river, from the point where it enters the cañon, winds tortuously from rock side to rock side, forming now and then rapids and falls, which preclude any attempt of ascending its waters in boats or canoes. The highest fall is six miles above its mouth ; the water presses there through rock cliffs not exceeding ten feet in width, and forming a cascade of over one hundred feet, reaches the bottom in a pond-like rock excavation. One mile below our night's camp is another fall about forty feet high and immediately at the camp a third one of fifteen to twenty feet. The width of the Pelouse river where we struck it is 200 feet, with gravelly bed and low banks, affording, at low water, easy fording places. Made to-day 12.5 miles.

June 8.—Started at $5\frac{3}{4}$ a. m. After crossing a little prairie ridge we reached, in five and eight-tenths miles from last night's camp, the top of a divide, (700 feet high,) which afforded a very good road, notwithstanding the deep, sandy portions of it. A rather steep descent of 300 feet brought us to a small ravine, with very rocky bottom, and following it for three and six-tenths miles, we struck the valley of "Cow or Cheranon creek." Ascending this valley for two and seven-tenths miles, we crossed the creek, which is, at the crossing place, thirty feet wide, and went to camp, having made eleven and four-tenths miles. Cow creek contains a small volume of water, which runs in a very irregular bed with respect to its width. The creek is scantily lined with brushwood intermixed only now and then with a grove of white thorn and cotton-wood trees. The road through the valley is much obstructed by little ravines and rocky elevations with steep ascents and descents, but a little work expended on this portion of the wagon road would render it perfectly easy. The bluffs enclosing the valley are rocky and bare of timber ; they vary very much in height and slope, and appear to become steeper and more rugged the nearer they approach the Pelouse river valley.

June 9.—After ascending the easterly bluff, (400 feet,) which was only effected by doubling teams, and even then with great exertions to the animals, we reached a level plateau bordered to the south by a prairie ridge of 300 feet in height. .Following this plateau in a northeasterly direction for two and a half miles, the road gains by an abrupt slope a second plateau, elevated 50 feet over the first one. Passing between two prairie ridges, which gradually sloped between Cow creek, we reached in four and a half miles a basaltic broken up country, which extends for many miles in an easterly direction ; but at the point where we crossed it, it has only a width of half a mile. Passing again between some isolated prairie ridges of a few hundred feet altitude, we struck, after three and a half miles, another basaltic bed, which extended to our camping ground, and was only occasionally interrupted by dry alkaline lakes which offered a firm and level road. Four and a half miles before reaching camp we passed a fresh water pond thickly covered with weeds and reeds, and surrounded by marshy ground. This pond and a little water-pool close by are the only watering places on this long and tiresome stretch. Coming to a creek which is said to run into "Big lake," but which, in my opinion, sinks into the ground or forms a branch of Cow creek, Major Lugenbeel turned off from Captain Kirkham's tracks and went up the creek, selecting his camping ground at a half mile distance ; made twenty and three-tenths miles. Captain Kirkham, as we had learned, had taken his road through "the cañon," and with hard work and much labor succeeded in getting his wagons through, though not without injury to them. This cañon is a narrow and winding valley of about two miles extent, filled with pedrigal rocks, which encumbered the passage of wagons considerably. Major Lugenbeel accepted, therefore, the offer of an old Spokane chief, "Big Star," to guide his train a different and better route. The creek we camped on has sufficient water in holes, and is bordered with brush at the camping ground ; but further below we saw big trees ; grass is good, but not over abundant. I had to-day twice an opportunity to see Pyramid peak, and took bearings to it.

June 10.—Resumed our march at 5½ a. m., and travelled for three and a half miles through level ravines, interrupted by two low divides. From the tops of the second divide we saw the "Big lake" before us with its pine-clad hills at the northeasterly end. A gentle slope of two miles brought us to the water's edge, and also to Captain Kirkham's road. The lake is surrounded by low hills ; some of them are rocky and precipitous; others again are of gentle grade ; both shores of the lake are entirely destitute of timber, and at few spots only brushwood can be found. Weeds and water plants cover the surface on both shores for about one hundred feet, and form a fresh, green garland round the quiet, glistening water sheet Travelling along the easterly shore for two miles we reached a point where, by a succession of rocky points, which extended far into the water, our road turned off from the shore, leading through little ravines and coulées which connected prettily with each other. Striking after one and a half mile the lake again, we reached again after three-quarters of a mile the northeasterly end of it, and turning short to the east we passed a luxuriant cotton-wood grove, and following up a wide bottom, which affords a good road in dry seasons, we crossed a little, marshy water-run, and camped near a spring which sends its icy waters to the lake ; made thirteen miles. The lake bears the name of "Big lake" in all the Indian languages ; it has a length of six miles and a width varying from one-half to three-quarters of a mile. The water is not fresh, and has a swampy taste ; but it may be that the plants and reeds on the shores may account for that. One rocky island, three quarters of a mile in length, is situated one mile from its southerly end. I took to-day one more bearing to Pyramid peak.

June 11.—Started 5.24 a. m. After ascending the small hill bordering our camping ground to the north, the road led for four and three-quarters miles over rolling prairie ground in the form of a valley, enclosed by distinct, though low, hill ridges ; striking then a marshy bottom which contains a little water-run, now and then enlarging to a pond, we followed it upwards for two and a quarter miles, and passing over a low divide, which is partially covered with pine timber, struck in one and a half mile Rock creek. This creek is fifteen feet wide, has a rocky bed, and is bordered by a luxuriant growth of large timber. The whole appearance of it denotes it as a mountain stream of some length. Lake Nechnichen, some sixteen miles northwest of the crossing point, receives its waters. In six and three-quarter miles, which were travelled over a gently undulating ground, enclosed by isolated prairie ridges, we entered an open pine-timbered forest, which is filled with innumerable little ponds and cut up by pedrigal rocks cropping out from the ground in great abundance. We passed over this kind of ground for three miles, and then striking the Spokane prairie went to camp on the head spring of Willow creek, having made twenty-two and four-tenth miles. Willow creek flows into Spokane river about four miles above the old Spokane house Its borders are well lined with cotton-wood, and the hills around are covered with scattered pines. After the spring freshet the creek sinks, rising only at intervals to the surface. The valley of Willow creek is well marked to the east by disconnected ridges, which are bare near the head spring ; but those nearer to the Spokane river are well timbered and of greater altitude. The most prominent peak is called Shkolsum or Bark mountain, and constitutes a good landmark. I saw plenty of snow yet on its tops, and to judge from that circumstance it must have a considerable elevation.

June 12.—Started at 5.40 a. m. Traversing a gentle prairie ridge we struck in three miles Muddy creek, a tributary of Willow creek. A marshy and timberless depression in the prairie forms a valley for this small water-run, and which requires a corduroy to allow wagons to pass. Five and a quarter miles further over ground of the same rolling character we reached a spring, and descended in one mile from there to the "Grand Coulée" by a very steep grade. The soil, which up to Muddy creek has shown no indication of great fertility, becomes, from that last mentioned point onward, very productive ; high and rich grass intermixed with flowers covers the hills, and groves of luxuriant trees are dotted all over the country. The "Grand Coulée" contains a rapid flowing creek of six to eight feet in width, which runs into Spokane river, and its side slopes are densely timbered to the water's edge. The ascent from the couleé is also very steep, but the pull is short. The next four miles of the road, which brought us to the brink of the highest terrace of the Spokane river valley, led across an undulating prairie country, frequently intercepted by strips of pine timber, which formed a communication with the large track of timber land lying to the east. The descent from the high plateau to the lower one is accomplished by following a long spur which strikes an intermediate gentle slope, which enabled us to gain the lower plateau by a curve. Following the foot line of the plateau slope, through open timber in a southwesterly direction for a mile and a quarter, we struck a sandy ravine, which brought us by a gentle grade in a mile and three-quarters to the water's edge of Spokane river, at a point where a private company has established a ferry. Having made eighteen miles we encamped on the left bank of the river where preparations were made for the crossing.

June 13, at 12½ p. m.—Everything having been ferried over we resumed the march. Ascending the river for two miles over rocky ground covered with open timber, we reached a point where the road leads up a narrow rocky ravine, in changing its course to a northerly direction. This bad place and the detour which it involves could have been avoided if the attempt had been made to ascend the plateau by the ravine which the trail follows, which would shorten the distance to "Walker's prairie" by four miles. The road which we followed gains the plateau by the aforesaid rocky ravine at the foot slope of a well marked rocky mountain, which forms the abrupt termination of a ridge bordering Walker's prairie to the east; keeping along that ridge we passed for six and a half miles on almost level ground through open timber, diversified occasionally by small prairies. Emerging on Walker's prairie we encamped on "Chimakane creek," having made nine and seven-tenths miles. This creek, which runs in a deeply cut bed, overflows Walker's prairie during the spring freshets to a considerable extent, and renders it swampy in many places. The prairie, which has a length of five miles, varies in width from a half to one mile, and is enclosed on both sides by high and well-timbered mountain ridges. Chimakane creek, which runs along the westerly edge of the prairie, is said to head in a pond which also forms the head waters of Mill creek, flowing in an opposite direction. The mouth of Chimakane creek, which empties into Spokane river, is seven miles below the ferry. Brushwood and trees of small size line its shores, furnishing sufficient wood for cooking purposes.

June 14.—Start at 5.36 a. m. After ascending the prairie for three miles the road enters the open pine timber, and gaining in two and a half miles an elevation of one hundred and twenty-five feet, we passed the mouth of a small valley, where the trail leading from the Cœur d'Alène country descends from the mountain to the Colville valley. After making a steep descent we struck, in one mile from the before-mentioned valley, the bottom through which Mill creek winds its way, and following it down close to the creek bed we ascended the plateau again in a mile and a quarter, and following its course over a gently undulating surface for three and a half miles crossed a clear mountain creek of ten feet in width, named "Clearwater." The next seven and three-fourth miles lead over a hilly country formed by the foot slopes of the easterly ridges, involving some steep grades. At a few places I noticed the rock cropping out from the ground, but only isolated huge boulders, allowing ample room to the wagons to wind their way along. Mill creek has been hidden from view since we crossed "Clearwater creek," but the direction of the valley is distinguishable by the bold ridges which mark its westerly limit. A steep descent brought us to a plateau running along Chal-mon-ets creek, but the heavy timber on it forced the road for three-fourths of a mile along the side hill before the regular crossing point could be gained. At this point the creek is twenty feet wide and two feet deep, but with the exception of this spot it is only an insignificant water run, which takes its origin in a lake about one mile above our camp.

The country has now assumed perfectly the appearance and character of the great mountain section lying to the east, and the suddenness of the change after a long journey over the mountainous plains makes the country appear to greater advantage than it really deserves. The soil of the hills and mountains is stony, and too sterile to produce even grass. The valley of the main creek is wet and swampy, and so much cut up by the winding streamlet that only patches of a few acres form a connected whole. Made to-day nineteen miles.

June 15.—Started at 5.36 a. m. After crossing the higher valley bottom, which is covered with open pine timber, we commenced a very gradual ascent to the opposite hill ridge, and passing in two and a half miles a small pond which lies in a hollow of the slope, we reached in one mile more a high hill, where I had a fine view of the westerly mountain ranges, which overtop each other in a terrace-like formation. The highest of these ridges is bare, but the lower ones are richly covered with pine timber. A large opening in the main ridge indicates that a large tributary of Mill creek, with a considerable valley, heads close up to the Columbia River valley, which itself cannot be at a greater distance than 15 miles. The highest point of the westerly ridges I estimated at 3,200 feet, and of the easterly ones at 2,800 feet. The elevation of the road at this point over Mill Creek valley is about 800 feet, which, with the exception of one point about one mile and a quarter to the north, is the highest altitude we gained. This latter spot is 200 feet higher, and from there the gradual descent to the valley is effected in five and a half miles, in which distance four little creeks are crossed, which are separated by low spurs. In gaining the valley the road skirted the open prairie, and we were compelled to keep along the foot slopes on account of the marshy character of the soil, involving thereby a detour of one and a half mile, and the crossing of two little muddy creeks, which sink as soon as they gain the open valley. A large valley opening towards the north, with well-marked, high rocky bluffs, indicates the course of a large arm of Mill creek. At the point where this stream has to be crossed we overtook the train of Captain Kirkham, which was delayed by the difficulty of the crossing, which is swimming and 20 yards wide. We went to camp, having made 11 miles.

June 16.—Having effected the crossing towards noon, we again took up the line of march, and reached, after crossing another creek, the first settlement of Colville valley, which of late has been abandoned on account of difficulty in raising crops. Passing then over a fine prairie, where a farm has been established, the road follows close to the creek bed for one mile and a quarter through open timber, over gently undulating ground, and then takes to a steep side hill, which closes in towards the creek, and has an extent of one mile and a quarter; after which distance the road gains again the valley plateau, and continuing on it for two and a half miles, surmounting only a few low spurs which run out from the main ridge, we went to camp one-half mile above a place called "Point of Rocks," having made 9.8 miles.

This "Point of Rocks" is a mountain slope formed by sliding rock of half a mile in length, the base of which is bordered by the swampy prairie bottom, which forced the command to work a practicable road over the slide itself. This place is the only difficult point on the whole route from Snake river, and during the dry season even that can be avoided. I notice one farm on each side of the creek, immediately beyond this place, and I am told that from here on the valley becomes more and more fit for cultivation. At this point the valley bottom is quite narrow, and not exceeding half a mile wide.

June 17.—The road ahead being practicable by 1 p. m., we started at 2 15 p. m. Travelling through open timber, after first passing round the first farm beyond the Point of Rocks, for three and three-quarter miles, and crossing in that distance five little mountain creeks, we debouched on a large open prairie of one mile and three-quarters in length. Several Indian farms are located here, and at its northerly end a number of log houses, belonging to Indians of the Colville tribe, are erected. We went to camp on a little creek, having made 6.8 miles. The mountain ridges enclosing the valley have a height of 800 to 1,000 feet, and are well covered with pine timber; their slopes are steep and in some places rocky.

June 18.—Started at 6 25 a. m. After crossing "Little Pend d'Oreille creek," which has a width of 15 yards and a depth of one foot and a half, the road leads for two and a quarter miles through open pine timber, crosses then a small but deeply cut creek, which required bridging, we passed between two small rocky cones, and reached in four and a quarter miles a place where Major Lugenbeel determined to establish a temporary depot. This latter section of the road is located along the foot hills of the main ridge, which now recedes very far. These hills are low, timberless, and covered with luxuriant grass, affording excellent pasturage, which makes them invaluable to the settlers of this valley. Mill Creek bottom widens considerably from the point where the road passed between the rocks, and gives ample room to farms on both sides of the creek. A quick ride, in company of Mr. McDonald, of the Hudson's Bay Company, brought me to Fort Colville, where I was kindly received by Mr. Blinkinshop, at the time in charge of the trading post. Returning the same evening to Major Lugenbeel's camp by the upper or wagon road, I took my leave, and towards night joined Captain Kirkham's train, which camped on the "Little Pend d'Oreille creek." Captain Kirkham, who had granted my request to return with his train, determined to travel as fast as possible with the view to meet you at Fort Walla-Walla, and accordingly he left his main train behind, pushing on with only two wagons, and a few dragoons of the escort. By travelling fast, and often during night time to avoid the heat of the day, we succeeded in reaching Snake river at 3 a. m. of the 23d. Captain Kirkham left for Walla-Walla the same morning, but owing to the weakness of my horse I could only start towards evening. On the morning of June 24 I met your camp on Dry creek, and obtaining leave to go to Fort Walla-Walla, I only rejoined your camp on Touchet river June 25. Spokane river is the largest stream crossed by the road; is 144.8 miles from Fort Walla-Walla; from there to the temporary camp of Major Lugenbeel it is 62.7 miles, making the whole distance 207.5 miles, as measured by my odometer. In accordance with Major Lugenbeel's request, I furnished him with a map of the road, upon which the main topographical features were marked, camping grounds stated, and distance given.

Very respectfully, your obedient servant,

P. M. ENGLE.

Lieut. JOHN MULLAN, *U. S. A., commanding Fort Walla-Walla and Fort Benton Military Road Expedition.*

CAMP ON THE BITTER ROOT, *June* 8, 1860.

Official copy for Office of Explorations and Surveys, War Department.

JOHN MULLAN,
1st Lieut. 2d Artillery, charge of Military Road.

C.

Report of reconnoissance of Pelouse river from last crossing of military road to crossing of Spokane trail, thence to Pyramid Peak and back, via a small tributary of Pelouse, made July 7–9, 1859.

SIR: In obedience to your order to examine the Palouse river and the adjacent country regarding its adaptability for the construction of a wagon road, I left the main camp near the Smokle creek on the morning of July 7, 1859, accompanied by Mr. Spangler and Donald McKay, our interpreter.

Over the first nine and one-half miles, to the mouth of the Oraytayoase, we passed rapidly, as this portion of the country was known to you by personal observation, and commenced our investigation beyond it. Having crossed the Pelouse about one-half mile above the junction of the Oraytayoase, at the same place where the military road crosses it for the last time, we found that 300 yards further up the stream the edges of the table lands bordering the valley of the river approached either bank very closely, forming a rocky defile one-half mile in length, sparsely timbered, even impassable for horses. We therefore left the river and ascended on a tolerably easy grade to the top of the table land on the right bank, which attains a height of about 350 feet. As far as we could see the country presented a high, slightly undulating prairie, destitute of timber, with here and there an isolated long, narrow ridge, from 100 to 250 feet above the general level. The principal depression therein, the valley of the Pelouse, was very distinctly marked by its rocky bluffs and scattered groves of timber. For a short distance we kept on level ground, thence descended on a pretty steep grade again into the valley of the river, character of which remained almost uniform for the whole distance we had an opportunity of examining it. It is one-fourth mile wide, much winding, its main direction E.NE. The river flows through it in a serpentine course, first touching one side and then the other. The banks on the outer side of these bends form invariably a more or less precipitous bluff of basaltic rock, from 200 to 400 feet high, according to the elevation of the table land itself, and are covered with scattered pines. The banks on the inner side are generally plateau, from ten to thirty feet above the surface of the water.. The current at and near these bends is swift, the depth of the water from two to three feet, the bottom big boulders. Above and below them the current is hardly perceptible and the water very deep. We found a trail, and followed it along the right bank of the river for seven and three-tenth miles, alternately passing steep side hills and level plateaus, to a point where it crosses and leaves the river, keeping in the valley of a small creek. We nooned in a little grove of tall pines, very conveniently situated near this crossing for the benefit of travellers. While our horses were enjoying the luxuriant bunch grass on the side hill, we feasted on service berries, wild currants, and gooseberries, which grow there and in many other places along the river in the greatest abundance. After a short rest Mr. Spangler and myself proceeded along the river. Donald McKay kept on the trail to find out its direction. In about four miles from this place we came to a considerable bend of the river, and saw in a northwardly direction the mouth of an immense cañon, with perpendicular basaltic walls 600 to 700 feet in height. A mile further we passed its entrance, and concluded, by the appearance of the country, that at some former period the river must have flown through it. At present it contains but little water, the bottom being marshy but in a few places, which were indicated by the growth of low bushes. Above this bend the topographical features of the valley slightly vary. The rocky bluffs become much higher—400 to 500 feet—and are of much greater extent—one mile or more long. The plateaus opposite are much narrower, and finally form only the lower portions of a gradual slope from the top of the table land to the river. The bottom land intervening widens, is extremely fertile, covered with tall grass, cottonwood groves, and wild currant bushes. Late in the evening Donald McKay joined us, bringing the intelligence that the Indian trail followed up an easy valley, near and almost parallel to the Pelouse, but much more direct. Three miles from the entrance of the basaltic cañon, or fifteen and one-half miles from the starting point of the examination, we encamped for the night in a rich bottom. So far existed no material difficulty to the construction of a wagon road. Immediately beyond our camp, though, was a rocky bluff, some 400 feet long, which could not be avoided on account of the marshy ground on the opposite side of the river. At daybreak, July 8, we continued still along the right bank as the easiest of the two. In three and one-half miles crossed a small creek coming from the northward. Half a mile further on we were compelled to cross over to the left bank on account of the rocky bluff on the right bank being impassable. Beyond this point we had to cross at every bend, the valley in general became very difficult, and altogether impracticable for a wagon road. The rocky bluffs abrupt—from 500 to 600 feet high—the bottoms either swampy or timbered, and the crossings deep, with very large boulders. In the same degree as the physical obstacles increased, the scenery became more beautiful and somewhat diversified at every bend of the valley. Once I tried to proceed on the top of the table land, but soon convinced myself that the innumerable ravines intersecting it in a short time would have broken down our horses and impeded our progress much more than the crossings.

So I again descended into the valley, repeatedly crossing the river, riding through fertile bottom lands, some one-half or even three-fourths of a square mile in extent, and occasionally climbing along rocky bluffs until I struck the eastern Spokane trail, which crosses the river at a point nearly fourteen miles distant from our camp of the previous night. We took a short rest to avoid the greatest heat of the day, and to give our horses an opportunity for grazing. Some Spokane Indians whom we found there encamped, behaved very friendly towards us, but could give no information in regard to the character of the country ahead of us. By all appearances, though, it seemed reasonable to suppose that the current of the river further up was rapid throughout, and the valley nearly approaching a cañon with but very small patches of bottom land. To reach the Pyramid Peak in the shortest time, we left the river entirely and ascended the steep face of the table land on the northern or right bank. For three and a third miles we rode on the trail in a northwardly direction, in a gradually ascending valley, formed by low prairie ridges. Gaining the top of a cross ridge at the end of that distance, we got a view of the Pyramid peak; as the trail did not tend towards it, we struck across the rolling prairie in a straight line for it. In seven miles we reached a bottom near the foot of the peak, wherein we found a series of cool springs, greatly to the delight of ourselves and animals. The sun was but a few degrees above the western horizon; we were much fatigued by the long march, and therefore rather disposed to delay the ascent until morning; but the delicious coolness of the atmosphere, succeeding almost instantaneously the heat of the day, inspired us with new energy, and we climbed up the steep slope of this rocky cone, that stretches its crown more than one thousand feet above its dingy neighbors. From the top we had a view of the whole country for eighty miles around us. The outlines of all objects were for a short while very clearly defined by the last rays of the setting sun. The Blue mountains, the high table land stretching from new Fort Walla-Walla to the Columbia and beyond it, the mountains around the Cœur d'Alène Mission and lake, and the Bitter Root mountains, were distinctly visible. Still the information regarding the particular object I was sent out for was by no means satisfactory. I merely convinced myself that the topography of the country, as developed by Mr. Sohon on various expeditions, was correct. The Pelouse, as far as I could see, followed an east and west course. The spurs of the Bitter Root mountain from which it proceeds were gently sloping, and densely wooded; pine timber in scattered groves reached from them to within four or five miles of the foot of the Pyramid Peak. The whole country enclosed by the above-mentioned mountain systems is rolling prairie, very much resembling a stormy sea. We encamped half way between the summit and the foot. The water we carried from the springs below, and collected dry fire-wood on the northern slope, which is covered with small brushes. The night was extremely pleasant, certainly much warmer than one might have expected. On the morning of July 9 we ascended once more, for the double purpose of taking some necessary compass bearings and enjoying the sunrise, which, beyond doubt, offers one of the most interesting spectacles. These objects accomplished, we retraced our steps to the Pelouse river, crossed it, kept on the Spokane trail, which, ascending the opposite table land, leads towards the south, and in three and a half miles struck an extensive and very rich bottom, the same which Donald McKay had examined for some four or five miles on the first day of our trip. We followed it down in as rapid a gait as our horses could endure; at intervals of from six to seven miles we found good springs, and small groves of cottonwood. Within five miles from the Pelouse the springs became more frequent, and finally formed a small running creek, which emptied into the Pelouse at the crossing of the trail, at the place where we nooned on July 7. There could be no route more desirable for a wagon road than this bottom offers, and as I have stated before that no material difficulties exist between it and the last crossing of the Pelouse near the mouth of the Oraytayoase, I would consider it as favorable a location for the military road as could be found in this region. This opinion, of course, holds good only as far as I went myself. Beyond the Spokane trail the bottom continues; whether it contains sufficient water, or whether the connexion with the valley of the Pelouse is easy, I cannot decide upon, although I suppose the difficulties would not be great. To avoid the rocky bluffs of the Pelouse valley below the junction of this bottom, we crossed over the table land, which would be a much better location for a wagon road. Towards evening we reached the last crossing of the Pelouse, where we found the whole command encamped.

I am, respectfully, your obedient servant,

THEODORE KOLLECKI,
Topographer.

Lieut. JOHN MULLAN, *U. S. A.*,
In charge of Fort Walla-Walla and Fort Benton Military Road.

Respectfully submitted to Captain A. A. Humphreys, United States topographical engineers, in charge of Office of Explorations and Surveys.

JOHN MULLAN,
1st Lieutenant 2d Artillery, Charge of Military Road.

CANTONMENT JORDAN, *February* 10, 1860.

D.

CŒUR D'ALÈNE MISSION, W. T., *August* 16, 1859.

CAPTAIN: I have the honor to report the arrival on this day of my party at this place. No accident worthy of note has occurred since the date of my last report, the 31st ultimo. Our camp at the time of my last writing was on the left bank of the St. Joseph's river, and at the place of crossing that stream. We crossed the St. Joseph's on the 1st and 2d instant. On the 5th we marched to the Cœur d'Alène river, about 11 miles. This distance contains some of the worst road yet passed. Our route extends up the Cœur d'Alène river along its left bank for about 11 miles. The total distance made since the date of my last report is about 29 miles. The greater part of the route since reaching the St. Joseph's river lies along the river valleys, and is subject to annual overflow. Some places were laid with corduroy before attempting to pass wagons over them, while many others showed by their softness that the waters had but recently left them. The road would be utterly impracticable for wagons during the time of high water, but for parties moving westward this would probably be no objection, as before they could pass the mountain snows the water would have subsided. The road has required much labor in its construction—cutting in side hills, and through forests, removing fallen trees, making bridges and corduroys. For a road of its class it may be fairly called a good one thus far. The St. Joseph's river as far as has been seen, and also the Cœur d'Alène up to this point, are narrow, (the former about 160 the latter 200 feet broad,) with scarcely perceptible current, and from 20 to 30 feet deep They may properly be called arms of the Cœur d'Alène lake, so far as the above description may apply to them. Above the Mission the Cœur d'Alène river assumes entirely a different character; it has a rapid current, and is easily forded. My observation has discovered nothing to induce me to change my opinion, as given in my last report, in regard to the dispoition of the Indians, nor anything to add in regard to them.

Lieutenant Lyon, acting assistant quartermaster, will forward by this express requisitions for cavalry boots. I have been led to ask for these, inasmuch as it is reduced now to an absolute certainty that we shall be exposed to snow during the march from this to Bitter Root river. Captain Kirkham, assistant quartermaster, has been requested to furnish the requisite number from Walla-Walla, in case those destined for us should not reach there in time. The health of the party continues good.

I am, very respectfully, your obedient servant,

J. L. WHITE,
1st Lieutenant 3d Artillery, Commanding Escort, &c.

Captain A. PLEASONTON,
Acting Assistant Adjutant General,
Headquarters Department of Oregon,
Fort Vancouver, W. T.

P. S.—*August* 17.—I open this to enclose a map, furnished, with Lieutenant Mullan's permission, by Mr. Engle. The red lines represent those surveyed by Lieutenant Mullan's party this year.

J. L. WHITE.

HEADQUARTERS DEPARTMENT OF OREGON,
Fort Vancouver, W. T., August 31, 1859.

Official.

A. PLEASONTON,
Captain 2d Dragoons, A. Assistant Adjutant General.

The following indorsement is upon the original:

HEADQUARTERS DEPARTMENT OF OREGON,
Fort Vancouver, September 1, 1859.

This report is respectfully submitted for the information of the general-in-chief, concerning the progress of the Fort Walla-Walla and Fort Benton wagon road expedition.

The Cœur d'Alène Mission is about 200 miles from Fort Walla-Walla. This expedition has been furnished everything that has been required for it, by the staff department of this command. Supplies are now en route for the Cœur d'Alène Mission to maintain the party through the winter.

WM. S. HARNEY,
Brigadier General, Commanding.

HEADQUARTERS OF THE ARMY, *October* 22, 1859.

Respectfully forwarded by direction of the general-in-chief.

H. L. SCOTT,
Lieutenant Colonel and Adjutant General.

Received at the Adjutant General's office October 24, 1859.

E.

CANTONMENT JORDAN, ST. REGIS BORGIA RIVER,
Bitter Root Mountains, W. T., January 2, 1860.

SIR: I have the honor to present the following report of my operations in the field during the months of September, October, November, and December, 1859, while engaged in running a line of levels over the wagon road from the Cœur d'Alène Mission to Cantonment Jordan. In obedience to your verbal instructions of September 14, I proceeded to the Cœur d'Alène Mission and commenced the level survey of the road on the 17th of the same month.

The water level of the springs at the western base of the Mission hill was chosen as the commencement point of the line, as it is the lowest point in the vicinity, and is also but a few inches above the level of the water in the Cœur d'Alène river at the Mission. The Cœur d'Alène river, which at the first crossing, one mile above the Mission, has a width of eighty feet, a current of a mile and a half an hour, and a mean depth of two feet, attains just below the Mission an average width of two hundred feet and a depth of at least fifteen feet, with a current that is so slow as to be imperceptible. This character is preserved from the Mission to the outlet of the river in the Cœur d'Alène lake, a distance of twenty-one miles in an air line, and twenty-seven by the windings of the river; for which distance the river, which here in fact is only an arm of the lake, is unquestionably navigable, and has but little if any fall. Six hundred feet north of the commencement point is a level plain, which extends to the northward three-fourths of a mile, to the base of the range of hills, which rise to a height of five hundred feet and form the northern boundary of the Mission valley, which has a length of three miles. This plain is covered with a fine growth of grass, which affords an abundance of pasturage for the bands of horses and cattle belonging to the missionaries and Indians. In the spring and autumn, however, it is mostly wet and marshy. Its eastern end is bounded by a high projecting spur of basaltic rock, which extends thence to the river, a distance of eight-tenths of a mile, and is the only rock of that formation which has been seen along the line of the road from the Mission to Cantonment Jordan. The hill on which the Mission stands, and along the base of which the wagon road passes, is eighty-two feet high, and presents a singular appearance, rising conically shaped out of a level plain, without being in any way connected with the mountain ranges on either side of the valley. Upon its summit are the buildings of the Mission, which consist of a large church and two houses for the dwellings of the priests and lay brothers, around which are scattered various tenements, lodges, &c , comprising the Indian village. At the northwest corner of the portico of the church building is placed bench mark No. 1, height above the commencement point, eighty-eight and twenty-nine hundredths feet. At the eastern base of the Mission hill is a gully thirteen feet deep and seventy-five feet wide, which drains the hills to the north and east, but which is dry during the summer months. At station three is the western fence of the enclosure known as the Mission field, which is used by the fathers as a farm, and extends to the east as far as the first crossing, and is bounded on the north by the basaltic spur before noticed and on the south by the river, having an extent of nine tenths of a mile in length by two-tenths wide. The first crossing of the river is made just below the point where the river leaves the spur, and being narrowed by a gravelly point which becomes an island at high water, runs swiftly over a gravelly bed, with an average depth of two feet and a width at the point of crossing of eighty feet. Should this ford be deepened, it is quite probable that the length of navigation on the river can be extended to near the Ten-mile Prairie. The general character of the banks of the Cœur d'Alène river at its crossings is very favorable for the purpose of bridging, being of a gravelly formation and perpendicular to the water's edge, and also not subject to overflow. The bed of the river throughout is also gravelly. A very good point for bridging the first crossing is just below the ford. The banks are eleven feet above the water level, and would be connected by a bridge of one hundred feet in length. The distance from the starting point to the first crossing is one and two-tenths mile. From the first crossing to the Ten-mile Prairie the road passes over three saddles, known as "the Four-mile Prairie Saddle," height one hundred and ninety-three feet; "the Seven-mile Prairie Saddle," height two hundred and sixty-three feet; and "the Ten-mile Prairie Saddle," height three hundred and seven feet above the base line. These saddles are portions of the long spurs extending from the ridges of mountains on the south to the river on the north, causing a large bend in the river, which otherwise runs in a remarkably direct line from its source in the mountains to the Mission. Both the first and second saddles are not at the lowest points in their respective spurs. The heavy growth of timber at these points effectually cut off the view of the surrounding country, and the want of time did not admit of their further examination. There are points of the spurs which are at least fifty feet lower than these saddles respectively. Between the first crossing and the Four-mile Prairie the road runs over a prairie for three-tenths of a mile, whence to the prairie it passes through a forest composed of cedar, hemlock, and balsam timber, varying from four inches to three feet in diameter. The Four-mile Prairie, four miles from the Mission, has an average width of two hundred and fifty feet, and is three-quarters of a mile long, bordered on the north by the river and on the south by a forest of large timber of the species before noted. The same character of growth covers the second saddle, and terminates at its eastern base. A large tributary of the river comes in from the north, it is supposed, nearly opposite to the eastern end of the Four-mile Prairie. This tributary is represented by Father Joset at the Mission as being the main fork of the river. He states that the Indians ascend it for some distance in their canoes. A glimpse of its valley has only as yet been obtained. The Seven-mile Prairie is a stretch of nearly a mile and a half, sparsely covered with grass and open timber. Its soil is gravelly. This prairie extends some distance back into the country to the south, and is the mouth of a large ravine which runs southwardly into the mountains. The many small gullies which run across it are evidently beds of streams in spring when the snows melt, although in summer perfectly dry. The hills which form the eastern boundary of this prairie rise in a peculiar manner from its surface. Instead of a curved line, as usual, joining the two plains of the side hills and the prairie, the place where they meet is a sharp and clearly-defined angle. This prairie is evidently the bed of a lake which has filled up the valley of this mountain ravine by its gravelly debris. The Ten-mile Prairie Saddle intervenes between the eastern end of this prairie and the Ten-mile or Mud Prairie. This saddle is the lowest point of its spur making down from the ridge on the south, and, like the others, is timbered from base to summit with a similar growth. The Ten-mile Prairie is about a mile in length, and from a quarter to three-eighths wide. It is wet and miry in spring from the melting snows, and in autumn from the rains; but at the time when passed over by our train was good and solid. A special survey is required from this to the Mission, in order to decide upon a definite location for a railroad line; but the wagon line is the shortest and most direct route either to the Mission or lake. Distance from the Mission to this prairie is nine miles; height of west end of prairie above the commencement part, eighty-three feet. From the Ten-mile Prairie to the second crossing the road passes through a forest of pine, hemlock, &c., and through a few small gravelly prairies, covered with grass and brush; the timber is generally small, being about fifty feet in height, and from six to twelve inches in diameter. The river at the second crossing has a width of seventy feet, and a mean depth of two feet. Its banks are gently sloping on the south and perpendicular and gravelly on the north; height above the river, six feet. One hundred feet below the crossing the river runs at the base of a spur, which last rises to a height of one hundred and fifty feet, an angle of inclination of forty degrees. Outcropping limestone rock here makes its appearance. From station sixty-four to station sixty-eight the road runs at the base of a steep, rocky ridge, (slope thirty degrees.) The ground to the south towards the river in this distance is marshy and covered with a luxuriant growth of long meadow grass. Further examination may show a better location for the wagon road on the left bank. From station sixty-eight to station seventy-two the road is through a forest of cedar, hemlock, and pine, from two to three feet in diameter, after which it passes into a prairie with open timber, which extends to within two hundred feet of the third crossing. The water level of the third crossing is one hundred and eighty-seven feet above the base line; its width is sixty-six feet, and mean depth two feet. From this crossing eastward the valley becomes narrow and the river exceedingly tortuous, flowing alternately at the base of the ridges that bound the valley on both sides, which rise up, with a slope of from thirty-five to seventy-five degrees, to a height of from three hundred to six hundred feet, where they slope off more gently to the tops of the high mountain ridges. For the heights and distances of the crossings, and different points along the line from here to the summit, I would refer to the subjoined table of heights and distances. From the third to the eleventh crossing, and from station one hundred and fifty-nine to the divide of the Cœur d'Alène mountains, the timber in the valley is of a heavier growth, and is more continuous than before; cedars from three to six feet in diameter frequently occur. The pine becomes abundant, attaining a growth of from two to four feet in diameter, and from one hundred and fifty to two hundred feet high. The hills and spurs of the mountain ridges are also covered with the pine, hemlock, and fir, which decrease in size in proportion to the elevation, until at the tops of the high ridges they are mere dwarfs, and the tops of the highest peaks are entirely bald. Many of these peaks bordering the valley have a height of from five to seven thousand feet above the level of the sea, the height of the range generally being from fifteen hundred to three thousand feet above the level of the sea, the height of the range generally being from fifteen hundred to three thousand feet above the level of the valley. A series of small prairies connected by strips of small pine timber begins near the eleventh crossing, and continues as far as station one hundred and fifty-nine, a distance of four and a half miles, only interrupted by two belts of timber, which together do not exceed a mile and a half in length. At the seventeenth crossing is, probably, the confluence of the north and south forks of the river. The south fork is again crossed just above its junction with the other branch. Height six hundred and twenty-five feet; distance from the Mission twenty-four and five-tenths miles. The first crossing of the north fork is one and one-tenth further to the eastward. Between this and the second crossing three small tributaries come in from the north, the valley of the stream becomes very narrow, and the ridges close up, forming a deep and narrow gorge, which at its mouth has a width of two hundred feet. This continues up to the ninth crossing, whence, to the twenty-seventh, a distance of two and seven-tenths miles, the river runs between the spurs which come down close, having only the width of the stream between them, with a narrow margin for the location of the wagon road. The character of the road as it runs over the small spurs and plateaus is indicated by the profile. The southern side of this ravine presents an almost unbroken wall, which slopes up from thirty to sixty degrees, with ledges of rock outcropping in many places.

The northern side, which is also steep, is intersected with four deep and narrow ravines, which extend to the summit of the mountain range, a distance of about two miles. The last or most eastern of these ravines, called "Johnson's Cut-off," will afford good grazing for emigrants and others passing over the line, grass being found in its bottom about a mile above its mouth, and also on the hills on both sides. From the twenty-seventh crossing the road rises up to a plateau, and passes over the Long Prairie Saddle, to avoid the Long Prairie, which is a marshy meadow, about one hundred and fifteen acres in area. One third of this area is covered with grass, and the remainder with clumps of bushes and strips of small timber This prairie is wet in early spring. The Long Prairie Saddle is twelve hundred and nineteen feet above the base line, and its distance from the Mission is thirty-one and three-quarter miles. After descending from this saddle to station four hundred and six, the road is located upon the foot slope of the ridges, which have a plateau like formation, and rise at an average rate of one hundred and eighteen and a half feet to the mile, in the direction of the road, until a summit is reached at station four hundred and forty-eight; thence it descends to the river bottom, and reaches the foot of the *divide* of the Cœur d'Alène mountains at the twenty-eighth crossing of the northern fork of the river. Distance from the Mission thirty-five miles and thirty-eight hundredths. Height fourteen hundred and fifty-two feet, or an average fall of forty-one and four hundredths feet per mile. The average width of the river is forty-four feet. The highest high-water mark seen was only three feet above its usual level. The average rise of the Cœur d'Alène river for the first ten miles is ten feet per mile, for the next fifteen miles thirty-five feet per mile, and for the remaining ten and thirty-eight hundredths miles is seventy-nine and seventy-six hundredths feet per mile. The summit of the divide of the Cœur d'Alène mountains at Sohon's Pass is four thousand nine hundred and thirty-two feet above the level of the sea, two thousand eight hundred and four and ninety-two hundredths feet above the base line, and thirteen hundred and fifty-three feet above the last crossing of the north fork, which last height is attained in an air-line distance of one and five hundredths of a mile. Its summit is gained by the wagon road by means of two long bends or curves, which are graded on and into the side of the mountain, the distance by the wagon road being one and seventy-two hundredths miles. The rock lies generally some four feet under the surface of the ground, but at the place known as the "Point of Rocks" blue limestone rock outcrops. From this place up to the summit the road runs on the natural surface of the ground. The descent of three hundred and ninety-two feet from the summit to the first crossing of the St. Regis Borgia river is made in a distance of seven-tenths of a mile, in a nearly direct line, over the natural surface of the mountain.

The divide of these mountains has had as careful an examination as the time at our disposal would allow, with a view to obtain as much correct information as possible for determining the best location and probable length of a tunnel for the passage of a railroad line. There are two passes to the north and east of Sohon's Pass. The first, two miles to the north, is higher than Sohon's Pass, and also wider from base to base. The second is some ten miles distant, at the north fork of the Cœur d'Alène. These require a special examination of their approaches to enable us to say anything definite in regard to them. At least a month should be spent in this region, so interesting to the topographer and the artist, to obtain at least an approximately correct sketch of the outlines of these passes and their approaches. Two miles to the south of Sohon's Pass the ridge of the mountains dividing the waters of the St. Regis Borgia from one of the small tributaries of the north fork of the Cœur d'Alène, which empties itself into that stream a mile or so above Long Prairie, becomes very narrow, sloping on the east at an angle of forty-five degrees, and to the west is nearly perpendicular for fifteen hundred feet. A tunnel could be made at this point of not more than one thousand feet in length, but its approach would involve such steep grades and heavy embankments across deep ravines as to be impracticable. The eastern base of the mountain in the valley of the St. Regis Borgia at this place, as determined by an observation with the aneroid barometer, is higher than the summit of Sohon's Pass It is therefore evident that to pierce the divide of these mountains either Sohon's Pass must be taken as the tunnelling point or a more favorable line be found through the mountain ranges to the north and east. The St. Regis Borgia river has its rise in a small bowl-like lake five hundred feet in diameter, carved by the hand of nature out of the steep rocky walls of one of the spurs of Stevens's Peak. After leaving this spring it tumbles down three hundred feet in a quarter of a mile through a narrow rocky channel, after which it flows through a valley a tenth of a mile in average width to the first crossing. The distance from the source to this point is about two miles, and the entire fall is twelve hundred and fifty feet. From the first to the second crossing, a distance of two and a half miles, the road passes over the foot slopes of the ridges on the right bank of the river, which have a general transverse slope of one in ten.

The descent from the first to the second crossing is five hundred and forty-nine feet. From the third to the sixth crossing, a distance of five-tenths of a mile, the valley is about one hundred feet wide, having only a narrow strip of ground between the river and the foot of the spurs. The rock is seen outcropping in many places, having a slope towards the river of from 40° to 70°. From the sixth to the seventh crossing, one and fifteen hundredths miles, the road passes for eight hundred feet at the base of the ridge on the left bank, whence for the remaining distance it passes over a plateau from sixty to one hundred and fifty wide, and four feet above the river bottom. Between the seventh and eighth crossings is a small prairie, five hundred feet long and two hundred and fifty feet wide, called the "Five-mile Prairie." The distance from the summit to the eighth crossing is five and sixty-five hundredths miles. From the eighth to the thirty-second crossing, eight and three-tenths miles, the valley is generally narrower, being from one hundred to two hundred feet wide. Wherever a prairie occurs the ridges and spurs recede and the valley becomes from two to six hundred feet in width. The principal prairies in this distance lie between the eighth and ninth, the ninth and tenth, and the thirtieth and thirty-first crossings. The valley between the last-mentioned crossings has a width of from five hundred to one thousand feet. The thirty-second crossing is at the entrance to the cañon of the St. Regis Borgia. which extends to the thirty-eighth crossing, a distance of one and a half mile. This cañon, so called, is merely a narrow and deep ravine or gorge, through which the river flows. The spurs which close up the valley have a slope of from 35° to 45°, and are covered with a growth of small timber; outcropping ledges of rock were only observed opposite to station seven hundred and ninety. The road is located on small plateaus at the base of the spurs for the most of the distance. From the fortieth to the forty-first crossing the road passes over rolling prairie ground, with belts of open timber. Here is the mouth of a large ravine, which comes in from the range of mountains on the north; the valley becomes wider and the river flows near the hills at the southern side of the valley. The stretches between the forty-second and forty-third and forty-fourth and forty-fifth crossings, are also the debouches of similar ravines. On these prairies will be found an abundant supply of grass during the spring and summer, although at the time of the survey, November 20 to 25, all these prairies were covered with snow to a depth of one and four-tenths feet The forty-first and forty-second crossings, and also the forty-third and forty-fourth crossings, are made in order to avoid two rocky spurs which run down to the water's edge at an angle of 35°. Here the blue limestone outcrops and presents a regularly stratified appearance; the thickness of the strata is from one to four feet, and dips from east to west at an angle of about 45°. The strata forming the spur between the forty-third and forty-fourth crossings has a dip of 60°, and the rock is of a whiter hue, more resembling marble. The forty-fifth and forty-sixth crossings are made to avoid a swampy flat occasioned by a stream coming from the north, which sinks into the ground instead of emptying directly into the river. From the forty-sixth or last crossing to Cantonment Jordan, a distance of one and ninety-six hundredths miles, we pass over a series of plateaus and rolling ground with prairies and belts of small timber. The character of the valley of the St. Regis Borgia is quite different from that of the Cœur d'Alène, rising in steps or plateaus of four or five feet each, instead of being a narrow bottom three or four feet above the water level. The timber both in the valley and on the sides of spurs and mountains is more open and smaller than the same character of growth on the western slope of the mountains. The pine is generally about fifty feet in height and from one to six inches in diameter, although occasionally trees are found much larger, the hemlock and fir predominating, while the cottonwood is seen lining the immediate banks of the river. Wherever the prairies already spoken of appear, the pines are smaller and evidently of a later growth, and the other timber becomes more open and scattering. The river itself has a more rapid descent than the Cœur d'Alène, as is shown by the following table, the average width being about thirty-eight feet:

	Distance, in miles.	Descent, in feet.	Average descent per mile.
From the 1st to the 2d crossing	2. 50	549	219. 60
From the 2d to the 9th crossing	3 57	312	87. 40
From the 9th to the 26th crossing	4. 48	282	62. 94
From the 26th to the 46th crossing	9. 05	354	39. 12
Total	19. 60	1,497	----------

The descent of the wagon road from the forty-sixth crossing of the St. Regis Borgia to Cantonment Jordan, a distance of one and ninety-six hundredths miles, is forty-five feet. The height of Cantonment Jordan above the base line at the Mission is eight hundred and seventy-one feet, and its distance from the Mission is fifty-nine and one-third miles, and from old Fort Walla-Walla, the initial point proper of the road, two hundred and eighty-eight miles. Height above the sea, two thousand nine hundred and ninety eight feet

I am, sir, very respectfully, your obedient servant,

W. W. JOHNSON.

Lieut. JOHN MULLAN, *U. S. A.*,
In charge Fort Walla-Walla and Fort Benton Wagon Road Exp'n.

Respectfully submitted to Captain A. A. Humphreys, United States topographical engineers, in charge Office Explorations and Surveys.

JOHN MULLAN,
Lieut. U. S. Army, in charge Military Wagon Road.

Place.	Height in feet above base.	Distance from the Cœur d'Alène Mission.
Mission hill	82	....
1st crossing Cœur d'Aléne river	10	1. 13
Four-mile Prairie Saddle	119	3. 25
Four-mile Prairie	40	4. 00
Seven-mile Prairie Saddle	263	5. 60
Seven-mile Prairie	138	6. 50
Ten-mile Prairie Saddle	307	7. 65
Ten-mile Prairie	83	9. 00
2d crossing Cœur d'Alène river	112	10. 65
72d station	153	11. 90
3d crossing Cœur d'Alène river	188	13. 27
4th.......do	193	13. 45
5th.......do	199	13. 65
6th.......do	206	14. 15
7th.......do	212	14. 30
8th.......do	217	14. 55
9th.......do	224	14. 80
10th......do	242	15. 52
11th......do	257	16. 25
12th......do	293	17. 38
13th......do	311	18. 00
Station 159	438	20. 80
14th crossing Cœur d'Alène river	488	22. 10
15th......do	521	22. 93
16th......do	566	23. 90
17th crossing and tributaries, station 204	584	24. 20
South Fork	606	24. 50
North Fork, 1st crossing	625	25. 05
Small stream	637	25. 25
2d crossing North Fork	645	25. 50
3d........do	653	25. 65
4th.......do	675	26. 00
5th.......do	707	26. 53
6th.......do	717	26. 75
7th.......do	720	26. 80
8th.......do	756	27. 37
9th.......do	826	28. 40
10th......do	838	28. 55
11th......do	854	28. 75
12th......do	862	28. 85
13th......do	868	28. 97
14th......do	871	29. 03
15th......do	892	29. 25
16th......do	907	29. 50
17th......do	915	29. 60
18th......do	936	29. 85
19th......do	948	30. 05
20th......do	956	30. 20
21st......do	968	30. 32
22d.......do	1,007	30. 55
23d.......do	1,010	30. 60
24th......do	1,023	30. 75
25th......do	1,028	30. 80
26th......do	1,042	31. 02
27th......do	1,045	31. 05
Long Prairie Saddle	1,219	31. 75
Station 406	1,167	32. 35
Station 448	1,507	35. 15
28th crossing North Fork	1,452	35. 35

TABLE—Continued.

Place.	Height in feet above base.	Distance from the Cœur d'Alène Mission.
Point of Rocks	2,669	36.80
Summit of Sohon's Pass	2,804	37.10
1st crossing St. Regis Borgia river	2,413	37.80
Small stream	2,136	38.95
2d crossing St. Regis Borgia river	1,864	40.30
3d.........do	1,833	40.55
4th.........do	1,822	40.65
5th.........do	1,805	40.75
6th.........do	1,799	40.80
Small stream	1,783	41.00
7th crossing St. Regis Borgia river	1,670	42.00
8th.........do	1,611	42.78
9th.........do	1,553	43.90
10th.........do	1,538	44.20
11th.........do	1,489	44.60
12th.........do	1,470	44.93
13th.........do	1,446	45.20
14th.........do	1,435	45.45
15th.........do	1,420	45.85
16th.........do	1,411	46.05
17th.........do	1,397	46.55
18th crossing St. Regis Borgia river; 1st stream	1,390	46.60
Do.........do.........2d stream	1,387	46.65
Do.........do.........3d stream	1,389	46.70
19th crossing St. Regis Borgia river	1,380	46.85
20th.........do	1,372	46.98
21st.........do	1,363	47.10
22d.........do	1,354	47.20
23d.........do	1,345	47.30
24th.........do	1,330	47.45
25th.........do	1,298	47.90
26th.........do	1,269	48.40
27th.........do	1,244	48.80
Small creek	1,248	48.90
28th crossing St. Regis Borgia river	1,232	49.10
29th.........do	1,215	49.47
30th.........do	1,207	49.65
31st.........do	1,174	50.45
32d.........do	1,152	51.10
33d.........do	1,148	51.15
34th.........do	1,136	51.30
35th.........do	1,132	51.40
36th.........do	1,125	51.80
37th.........do	1,100	52.35
38th.........do	1,093	52.55
39th.........do	1,080	52.85
40th.........do	1,074	53.05
Stream	1,054	54.30
Small creek	1,012	55.10
Do	1,011	55.15
41st crossing St. Regis Borgia river	978	55.30
42d.........do	975	55.40
43d.........do	945	56.25
44th.........do	940	56.50
45th.........do	922	57.20
46th.........do	916	57.40
Cantonment Jordan	881	59.33

F.

CANTONMENT JORDAN, ST. REGIS BORGIA RIVER,
Bitter Root Mountains, January 3, 1860.

December 27.—SIR: I have the honor to state that I left the Mission at eleven o'clock a. m., in company with two Indians, on our way to Cantonment Jordan, with about twelve pounds weight to the man, and camped on Mud or Ten-mile Prairie. The snow at this point is two feet and five inches in depth. Travelling on snow-shoes.

December 28.—Left Mud Prairie at six o'clock a. m , and camped at Cedar Swamp. Good travelling. Snow same as the day before. Distance 25½ miles from Mission.

December 29.—Got under way at seven o'clock a. m. Good travelling. Snow three feet in depth, with a hard crust. Camped at Long Prairie; distance from Mission 32⅓ miles.

December 30.—Left camp at six o'clock a. m.; crossed the divide, and camped at the Log House, six miles east of the summit.

From the Cœur d'Alène river to the summit the depth of snow is four feet and seven inches; from the summit to St. Francis Borgia river the snow is five feet deep; from this point to the Log House it is four feet and six inches.

December 31.—Got under way at seven o'clock a. m. Good travelling. The road had been opened by Lieutenant Lyon's men, who were sledding their wagons to camp. Snow two to three feet deep on an average. None of the crossings between the Mission and the divide are as yet frozen. Total distance travelled sixty miles on snow-shoes.

I am your obedient servant,

P. TOOHILL.

Lieut. JOHN MULLAN,
Charge of Military Road.

CANTONMENT JORDAN,
St. Regis Borgia River, January 4, 1860.

SIR: The above report of my expressman, P. Toohill, of his trip from the Cœur d'Alène Mission to our winter camp, made in December, gives the exact depths of snow along the line in the month of December, 1859, and may, with other data, serve to form some definite conclusion regarding the actual depth of snow on this route during the winter season.

I shall have a report, also, on the depth in January, February, and March, made and submitted to you.

I am, sir, respectfully, your obedient servant,

JOHN MULLAN,
1st Lieutenant 2d Artillery, charge of Military Road.

Capt. A. A. HUMPHREYS,
United States Topographical Engineers.

Report on the practicability of a railroad along Columbia river, from the Dalles to Umatilla river. 1859. *C. P. Howard, C. E.*

CANTONMENT JORDAN,
Bitter Root Mountains, January 5, 1860.

SIR: In accordance with your letter of instructions, I left the Dalles on May 16, 1859, for the purpose of examining into the practicability of a railroad line, along the Columbia and Snake rivers, from the Dalles to the mouth of the Pelouse river.

Before entering into details, it would be, perhaps, better to say something of the general character of the route, and more especially of the question of grades, which in most lines of railway is the subject of chiefest difficulty. A glance at the accompanying profile, together with the following considerations, will show that if, under any circumstances, a railroad is practicable along the Columbia and Snake rivers, the problem of grades is of easy solution. A large portion of the line is on the broad flats between the bluffs and the river, where there is no engineering difficulty to be surmounted. A considerable part also—and such sections will be examined in greatest detail—lies along the face of steep rock bluffs, with the river laving their base, and where, by a variation of a few hundred feet to the right or left, the line can be elevated or depressed the greater part of the distance; here it is sufficiently evident that the question is not one of grades. Still another port on is on rolling plateau which sometimes continue for miles, with very gentle ascents and descents, and then break off suddenly, with sheer precipices of from fifty to one hundred feet in height, compelling the engineer to take the steep side hills beyond; and these sections are the only ones on which any difficulty from grades can be apprehended. Further investigation will show that even here the difficulty is not great

To afford a cheap line of railway from the Dalles to the Umatilla river—beyond which point I will not enter into detailed report, from being incapacitated for field-work by sickness during the remainder of the exploration—the amount of curvature must of necessity be considerable, but much less than those unacquainted with the peculiar features of the country might suppose. In almost all cases where large ravines occur the bluffs recede from the river, and broad flats intervene; and where the precipitous rock cliffs come down to the water, a straight line can often pass along their face for half a mile without leaving the surface more than a few feet. There are many points along the line where a large amount of rock cutting and embankment would be necessary, and probably in several places short tunnels would be required; but the average cost of the line would not be great if curves, of one thousand feet radius as a minimum, should be adopted where the nature of the ground requires them.

Regarding the means and manner of operation during the exploration, the instruments used were a compass, spirit-level, and odometer; the latter attached to a small wheel, fitted with axle and shafts, and rolled along the line as surveyed. Mr. Sohon was compass-man and topographer from the Dalles to Mud Spring; thence Mr. Johnson took his place to a point some ten miles west of John Day's river, and the remainder of the line was surveyed by Mr. De Lacy. The level was used by myself until we had nearly reached the Umatilla, where sickness forced me to leave the party, Mr. Johnson taking my place as leveller, and the charge of the party devolving on Mr. De Lacy, who completed the special exploration. The map of the line was made by Mr. Sohon so far as he surveyed it, and the rest by Mr. De Lacy; the profile by myself. In the following report, as in that of Mr. De Lacey, the stations referred to are taken from the level-book, and are identical on map and profile.

The high-water mark, as given on the profile, was obtained from actual observation of the greatest height of water of the Columbia river in 1859, when, I am assured, it was higher than at any previous time in the knowledge of the white residents of the vicinity; and it may safely be predicated that a line of railway, located a few feet above this mark as given, would be safe from any encroachment from the Columbia.

It is necessary to state that whilst the profile was made entirely from the level notes, the level line as run was not, in many instances, exactly over the ground which was chosen as the best line of location; but being limited in time, as well as means, during the survey, and there having been no previous examination of the ground, I was forced to content myself with merely noting in the field book the variation between the line as run and that afterwards selected. The field-book, therefore, will frequently show the data used in this report where the profile does not.

The initial station of the survey was taken at a point on the Walla-Walla road, two miles east of Fort Dalles, and a bench-mark made on the root of a small oak tree near this station was taken at one hundred feet above the base line.

From station 4 to Five-mile creek, distance two miles, the line rises 45 feet to the point selected as the best bridge crossing. About one mile of embankment six feet high, and a quarter of a mile of rock cutting five feet deep, necessary. Small culvert, six feet span, necessary at Three-mile creek. Length of bridge across Five-mile creek 140 feet; good natural rock abutments at about the same level, and 50 feet above water.

From Five-mile creek to 27, distance one and a half mile, the fall is five feet. Between these points the ground is much rougher; about three-quarters of a mile of rock cutting 8 feet deep, and half a mile embankment 10 feet high, necessary.

From 27 to 32, distance one mile, line rises 10 feet. Work here is mostly in sandy earth. About half a mile of earth cut 15 feet deep, and half a mile embankment 15 feet high, will be required.

From 32 to 40, distance three miles, rise 10 feet. Between these points the quantity of work necessary is estimated as about equal to seven-eighths of a mile of rock cut 10 feet deep, and half a mile of embankment 30 feet high.

From 40 to 43 the section is much more difficult than any yet considered. These points are on the same level, and their distance apart is one and three-eighths of a mile. Between them will be necessary half a mile of embankment 20 feet high, which extends for some 800 feet across a shallow corner of the river; then a tunnel of quarter of a mile, about; and thence to station 43 the line passes for some 600 feet over another and much deeper corner of the river, where a bridge will probably be required. The exact depth of the river at this point could not be ascertained; but at high water it must be at least 20 feet, and the current, for a part of this distance, is quite rapid.

From 43 to 53, distance two and a quarter miles, rise 77 feet, grade 34 feet per mile, there will be necessary one mile of embankment 10 feet high. Between 44 and 53 the cliff makes a sharp bend, requiring a curve of 2,000 feet radius, and angular deflection of 80°.

From 53 to 54, distance 0.44 mile, line level; rock cut averaging 10 feet for the whole distance. This is an elevated plateau which breaks off suddenly at 54, with a fall of some 70 feet.

Between 54 and 55 is another difficult section. The line leaves 54, passing along a very steep side hill for about a quarter of a mile, then curves for some 30°, with radius of 1,000 feet, and passes, on a straight line, through a rock bluff, with a tunnel of about 1,000 feet; then, with a curve of larger radius, and some 90° of curvature, it keeps the rising ground behind the Des Chutes village, and passes on a straight line across the Des Chutes river to station 55. For the first 1.40 mile the fall is 42 feet, grade 30 feet per mile; thence the line crosses the river on a level, and 15 feet above high-water mark, to station 55. Length of bridge across Des Chutes river, 480. Good abutment and pier foundations. Whole length of embankment at both ends of bridge, 900 feet; height, 15 feet. The whole distance from 54 to 55, on this location, is 1.85 mile; but, by a mistake in the odometer readings, which was not discovered until after the profile was finished, the distance, as it appears on the profile, is 0.67 mile too short. This will be noticed on the profile as well.

From 55 to 61, distance about one and a half mile, rise 7 feet; work about equal to earth cut six feet deep the whole distance. From 61 to 62, distance rather more than one mile, fall 16 feet, the grade line scarcely leaves the surface more than a foot either way. Material all sandy earth.

From 62 to Mud Spring, distance two miles, rise 34 feet, one mile embankment 10 feet high, and quarter mile earth cut six feet deep, necessary. A culvert of six feet span required at Mud Spring.

From Mud Spring to 79 the profile of the selected line of location is entirely different from the level line as run. Between Mud Spring and 74, distance two and a quarter miles, fall 34 feet. The line for the first mile is along the slope of very precipitous rocky bluffs; work for this distance, probably half a mile of rock cut 20 feet deep. For the remaining one and a quarter mile the work is very light and nearly all in earth.

Between 74 and 79, distance seven miles, rise 68 feet. The line here lies mostly along the level or gently rolling river flats, where both excavation and embankment would be slight, and the material nearly all earth. For a distance of 2,000 feet back of 78, however, the line is near the base and along the surface of nearly perpendicular rock bluffs rising from the river. Here work would be required equivalent to rock cutting some 50 feet deep; but even should a tunnel of 2,000 feet be required, it would be much preferable to the heavy grades of the line as run and indicated on the profile. About three-quarters of a mile back of 79 is a ravine where a culvert of 12 feet span would be required; also two small ones of four feet span between 76 and 78.

Between 79 and 82, distance one mile, is John Day's river. The line falls 16 feet in the first half mile, crossing the river on a level, which continues to 82. Length of bridge required, 450 feet. Foundation for east abutment solid; both west abutment and pier foundations would probably require piling. Embankment on west side, for quarter of a mile, averaging 20 feet in height; on east very little needed.

From 82 to 90, distance three and a quarter miles, line continues level. No engineering difficulty between these points. 1,000 feet of rock cutting 20 feet deep, and quarter of a mile of embankment 20 feet high, would be about equivalent to the work needed. Culvert of four feet span between 86 and 88.

From 90 to 92, distance two miles, fall 20 feet. 1,000 feet of rock cutting six feet deep required; no embankment; one culvert of five feet span.

Between 92 and 95, distance two miles, rise 60 feet, there are two very pricipitous rock bluffs of 100 and 80 feet in height, respectively. The line passes about 15 feet above the base and along the face of the first for some 1,500 feet. Here trestle-work would probably be most economical. The second is about two-thirds of a mile back of 95, and by cutting through this for 300 feet, with average depth of 15 feet, the line would pass with very little more work up to 98.

Between 95 and 108, distance a little less than five miles, fall 37 feet, three-quarters of a mile of rock cutting 10 feet deep, half mile earth cutting 20 feet deep, and one mile embankment averaging 20 feet in height, will be necessary.

From 108 to 120, distance four and a quarter miles, fall 7 feet; about three-quarters of a mile earth cut 10 feet deep, and three quarters of a mile embankment 8 feet high.

From 120 to 121, distance two miles, fall 12 feet, a very special survey is needed. The line passes across a shallow corner of the river for some 800 feet, and about one mile is along the face of very tortuous, broken, rocky bluffs, 60 feet high.

From 121 to 126, distance one and a quarter mile, rise 20 feet, half a mile earth cut 10 feet deep, and half a mile embankment 10 feet high, necessary.

From 126 to 138, distance three and three-quarter miles; grade line level; earth cut one and a half mile, 6 feet in depth; embankment almost nothing.

From 138 to 175, distance 11 miles; the fall here is 17 feet in the first four miles; then the line rises 33 feet in the next six miles, crossing Willow creek on a level and 13 feet above its banks, to 175; for the first 9¼ miles the grade line does not leave the surface at any point more than 10 feet, and for this distance two miles embankment averaging eight feet in height, and one mile earth cut 10 feet deep, is about the amount of work required. About one mile back of Willow creek the line passes along a rocky hill-side, and from this point to the creek one-half mile of rock cut 10 feet deep, and one-quarter mile embankment 15 feet high, is necessary. Length of bridge across Willow creek 330 feet; abutment foundations solid; pier foundation will probably require piling, one-quarter mile embankment averaging six feet in height between Willow creek and station 175.

From 175 to 196, distance is 4¾ miles, fall 15 feet. In the first half mile there is required one-quarter mile rock cut eight feet deep; thence up to 196 the grade line does not anywhere vary more than 10 feet from the ground. For this distance, one mile earth cut six feet deep, and one-half mile embankment six feet high, is necessary.

From 196 to 211, distance 2¾ miles, rise 32 feet; embankment averaging 15 feet high for one mile.

From 211 to 215, distance half mile, rise 11 feet; excavation and embankment almost nothing.

From 215 to 222, distance 2¼ miles, half mile of rock cut eight feet deep. The fall between these stations is 26 feet.

From 222 to 257, distance 6¾ miles, rise 35 feet. There is necessary half a mile embankment 31 feet high, two miles embankment eight feet high, and three-quarters of a mile earth cutting 10 feet deep.

Between 257 and 258, distance 8¼ miles, the fall is seven feet. The line here runs straight across a bend of the river on a low, gently rolling flat. There is nowhere cut or bank of more than 20 feet, and the material is all sandy earth. Over this section the level did not follow the selected line of location.

From 258 to 273, distance 1¼ mile, rise 37 feet; about half a mile embankment 10 feet high necessary.

Between 273 and 325, in the Umatilla river; the distance here is four miles, and the line level; half a mile embankment 20 feet high, and quarter of a mile earth cut five feet deep, necessary. Length of bridge across the Umatilla 340 feet. Solid foundation for abutment and most probably for piers; the banks down to the water's edge being remarkably firm.

The section of the line from the Dalles to the Des Chutes river, a distance of 13 miles, presents far more of engineering difficulty than any portion of the same length up to the Umatilla. For most of this distance the bluffs are generally very precipitous and near to the river; the peculiar basaltic formation, in some places, rendering anything approaching to an exact railroad location almost impossible, except from a very detailed survey; whilst in others the position for a railroad is so strongly marked that a very approximate location can be made without an instrument; it is on connecting these last-mentioned portions with easy grades and great economy that the whole real difficulty lies. Between the Dalles and the Des Chutes river, the ground which, of the whole distance to the Umatillah, may be considered as most unfavorable for a railroad, amounts to nearly eight miles; requiring in this distance probably 2,200 feet of tunnelling, and 1,000 feet of bridging; whilst for the remaining 80 miles the sections presenting like difficulties amount to only 4½ miles. With the exception of these 12½ miles, the estimate of the work required is quite approximate, most of the line being on broad sandy flats or gently sloping side hills.

The rock is all basaltic, but, being much seamed and broken, is generally favorable for blasting. It is also well suited for rubble-stone masonry and ballasting, for which purposes material in abundance can be obtained from the debris at the base of the cliffs, which at some points has its natural slope for some 100 feet of vertical height.

Clean, sharp sand of the finest quality for building purposes is in abundance at many points.

There is no timber along the line; a few small willows, at long intervals near the margin of the river, being the only wood; pine timber, however, can be obtained on the Yakima, 12 miles from the Columbia, and also at the mouths of the Spokane and Pisquouse rivers, which may be rafted down the main Columbia. Pines and cedars are also found along the Clearwater, which can be brought down the Snake and Columbia rivers.

I am, sir, most respectfully, your obedient servant,

C. R. HOWARD,
Engineer of Military Wagon Road Expedition.

Lieutenant JNO. MULLAN, *U. S. A.*,
Commanding Military Wagon Road Expedition
From Walla-Walla to Fort Benton.

H.

CANTONMENT JORDAN,
Bitter Root Mountains, January 5, 1860.

SIR: In consequence of the illness of Mr. Howard, who was obliged to retire from the field, I was placed by you in charge of his party, and continued the survey of the Columbia and Snake rivers from the mouth of the Umatilla river to the mouth of the Pelouse.

The work was completed on the 31st of July, after which the party took up the line of march for your camp, which we reached on the 7th of August, on the Cœur d' Alène river. During these operations the *personnel* of the party remained the same as when left by Mr. Howard. The level was run by Mr. Johnson and the topography of the line was taken by myself. The instruments used were odometer, engineer's level, and prismatic compass.

Previous to entering on the details of the survey, which I shall give as far as the time now at my disposal will admit, I would remark that on the portion surveyed by me, as on that previously surveyed by Mr. Howard, the question of grades, often so difficult a problem to solve may be said not to exist; that is, they are so light as to cause no difficulty whatever. It is simply a question of curves and construction; or, in other words, to put in curves which will require the minimum amount of construction. This will be peculiarly the case at some of the points of basaltic rock which occur on Snake river, most of which project in the form of ledges, which can be easily taken advantage of for the road-bed.

The line of the river was followed throughout. No attempt was made at an absolute railroad location, but the best was done that the time and means at our disposal would permit. The profile and map will show the character of the country, and the information gained will serve to guide future engineers to the most difficult points for more detailed examination.

The stations referred to in this report are the level stations marked on the profile, and commence at No. 325 on the right bank of the Umatilla river. At this point the line is distant about half a mile from the Columbia, but reaches it again in a mile and a half with a descending grade of six feet per mile, passing over a sandy earthen flat, and then takes along the river for three miles more to station 363, on top of a rock ledge and along a gentle sand slope, which admits of a grade of 10 feet per mile. At 363 the line takes a steep, rocky side hill, with occasional perpendicular ledges, for three and a half miles along the promontory called the "Monumental Rocks" to station 387, where they cease. The grade in this distance will not exceed five feet per mile, but will probably require, say 1,000, yards of rock cut ten feet deep. The rest will be earth excavation, and curves will be from two to three miles radius. There is one large ravine 300 feet wide, through which flows a small spring branch. This will require embankment, and a culvert of six feet span.

From station 387 to station 447, distance 5.4 miles, the ground falls 10 feet, passing over a rolling, sandy country, where there would be about one mile of embankment six feet high, and the retaining distance cut; average depth, three feet.

From station 447 to station 496, distance 4.18 miles, the ground rises 15 feet, and will require very little work, except in the last mile, where there will be about 3,400 feet of embankment three feet deep, and 800 feet rock cut 30 feet deep through a rocky point jutting into the river. The level line between these stations did not follow the selected line of location.

From station 496 for about two miles the grade will be very light, almost nothing. There will be 200 feet rock cut eight feet deep, and about 3,000 feet of embankment; average height, five feet. A better location could, however, be probably found by throwing the line nearer to the foot of the hills. Thence to the Walla-Walla river station 576, a distance of three miles, the country is rocky and broken. Points of rock jut into the river, and ledges of rock occur cut up by ravines. The line will pass at the foot of these, with a very light grade. There will be about one-third of the distance embankment 10 feet high and two-thirds rock cut. A special and detailed examination will be required in order to locate the line or make an estimate of the amount of work.

On the left bank of the Walla-Walla an embankment of 1,500 feet long and 10 feet high would be required, and a similar one on the right bank for a mile. A bridge, crossing about 100 yards above the mouth of the river, of 550 feet in length, would be required. Both banks are of sand, there being no rock immediately on them, but that material can be obtained in abundance from the neighboring cliffs. The river is very rapid, and fordable in low water.

From station 576 to station 587, or between Walla-Walla and Snake rivers, a distance of $10\frac{1}{4}$ miles, the whole country is a sandy flat, almost level. There are but two elevations on it, neither of which is over 20 feet in height, and both can be avoided, as well as two or three gullies which occur, by throwing the line further from the river towards the low range of sand-hills which border it at the general distance of three-fourths of a mile.

On each bank of Snake river an embankment of about 2,400 feet in length and 10 feet in height will be required. A draw-bridge of 1,500 feet will span the river at a point about a quarter of a mile above its mouth. The river is about 16 feet deep at this point in high water.

Curves of 1,500 feet radius will probably be necessary to approach the bridge on either side, particularly on the right bank, when a range of hills about 100 feet high, following down the Columbia, come close on the bank of Snake river about one-half a mile above its mouth. This river was crossed by the survey at this point, and the right bank is the one always spoken of below.

From station 589 to station 598, (right bank of Snake river,) distance four miles, the line runs along the foot slopes of a range of sand and gravel hills, having the same general grade as the river, or not more than four feet per mile. There will be two embankments with drains across the mouths of ravines, one of 300 feet length and three feet depth, and another of 600 feet and same depth. Just beyond this station is a rocky point, which will require 352 feet of cut of 35 feet depth. For the next mile to station 601 the line runs on a flat, following the natural surface of the ground, and crossing a slough which will require a bridge of 75 feet length.

From station 601 to station 632, distance nine miles, the line, with the exception of about one-half mile when it runs on a flat, takes the sides of gravelly hills with slopes varying from 20° to 45°; no grade in this distance will exceed 10 feet per mile.

The mouths of several ravines are passed; three of 50 feet wide and three feet deep; one of 30 feet wide and 10 feet deep; and one large one of 500 feet embankment, 25 feet deep, with culvert of six feet span. None have water in them except in the spring. The rock cutting between these stations will not exceed 40 linear yards. All the rest will be earth excavation. Along this portion of the line the river makes a very large bend northward.

From station 632 to station 635, distance one mile and three-eighths, the line runs for one-half a mile along the steep side of a high rolling plateau, with a ledge of rock next the river, until a slough comes in which makes a large bend towards the west and cuts off a large gravel island.

Beyond the slough is a high rolling flat, extending for a mile to the foot of what is called "Anchor cañon," where the river is enclosed on either side by perpendicular ledges of basaltic rock rising directly from the water. The cañon extends from about 646 to about a quarter of a mile beyond station 663, where there is a deep ravine and where it ends.

There are two distinct ledges of rock continuing all the way to this ravine; the lower one occasionally interrupted and narrow, and from 30 to 50 feet above the water.

These ledges are remarkably straight. All of this section, from station 632 to 646, will need a very special and detailed examination to ascertain the best and most economical location. The most obvious would seem to be to cross the above-mentioned island, which would require two bridges of about 300 feet in length, and embankment of about 1,500 feet in length and 30 feet in height, taking advantage of the rolling flat, beyond which is much higher than the island, to obtain the requisite grade, (which in no case would exceed 30 feet per mile,) and thus pass the cañon on the upper or lower ledge as may be found most advisable. There will be about two and one-quarter miles of heavy excavation and embankment, and about the same amount of rock cutting 20 feet deep. Just beyond station 663 is a ravine about 500 feet broad, which will have to be passed, and a point of rock requiring a cut of 200 feet long, 40 feet deep. From this point to station 721, distance seven and one-half miles, the line will run generally on the sides of gravel hills with slopes of from 20° to 40°, with about a mile of rolling plateau. There will be some seven or eight large ravines requiring embankments of from 100 to 400 feet in length and an average depth of 10 feet, with culverts, and numerous small gullies only requiring drains. There will be about 4,000 feet of rock cut with an average depth of six feet and about 500 feet of retaining wall. Grades very light, not more than 10 feet per mile.

From station 721 to station 738, distance two miles, will require a detailed examination to determine the best location. About one-half a mile consists of a rolling plateau cut up by ravines on top, with two rock ledges at the base and part (one and a half mile) of a bluff, the face of which is composed of alternate broken ledges of rock and steep earthen slopes. The line on the profile gives the levels on top where the instrument was enabled to run, above which again are other perpendicular ledges or cliffs over 200 feet high. A line can be obtained along the face of this bluff by taking advantage of the ledges of rock, and the use of retaining walls.

It is impossible to give any definite idea of the amount of work, but most of it will be rock cutting. A tunnel would not be needed, and the grade would not exceed 20 feet.

From station 738 to station 742, distance one and three-quarters mile, the line runs generally along side hills with slopes of from 20° to 30°, where the grade will be light. In this distance there are four ravines, embankments from 60 to 200 feet in length and average depth of 25 feet. Culverts four feet span.

From station 742 to the Pine Tree rapids, distance three and one-quarter miles, the river is bordered by steep hills cut up by ravines. The lower portion of these hills consists of broken ledges of basaltic rock, points of which jut into the river, and have elevations of from 30 to 75 feet above the river. There are five ravines which will require bridges or embankments of from 100 to 350 feet, and two which are 900 feet broad. There are rock ledges on both sides of all these ravines. The grades would be light, but at least one-half of the work would be in rock cutting, the remainder earth excavation and embankment.

From Pine Tree rapids to station 757, distance three miles, the line would run along a broken rocky ledge with a descending grade of 20 feet per mile. This last section will have about 2,000 yards of rock cutting 10 feet deep. The rest being earth excavation on side slope of one and one-half to one. Three ravines are to be passed, 275, 300, and 350 feet broad, on the line. This section will require a very particular examination, being probably the most difficult on the whole river. From station 757 to station 758, distance one and one-quarter mile, the line of location selected does not follow the line exhibited on the profile, but takes the bank of the river with an ascending grade of 15 feet. Two large ravines will be passed requiring culverts of 10 feet span. From station 778 to station 783, distance two miles, the line has a descending grade of 10 feet per mile, and presents no difficulty as it follows the curve of the river bank on side slopes of earth, varying from 15° to 45°, and intersected by only two ravines of any size, requiring culverts of 10 feet span.

From station 783 to station 797, distance 2½ miles, the line can be located either according to the line shown on the profile or on either side of it, as may seem on minute examination the best line to approach a basaltic cliff which juts out into the river. There is no difficulty whatever on any of these lines, and no rock cutting.

At station 797 the cliff commences and continues for about a mile. There are three ledges of rock running parallel to the water but not continuous. Curves of 1,500 feet radius would be required at station 799, and the line can run around the face of the cliff on the lower ledge. There will be rock cutting for a mile with an average depth of 15 feet. This point will require a very particular examination. The grade will not exceed 10 feet per mile. Within the next half mile to station 799 there are three large ravines which will require embankments of from 200 to 300 feet in length and 20 feet in depth—culverts of 10 feet span. In the above-mentioned cliff the first ledge rises about 50 feet above the water, has an average breadth of 30 feet, and would be the best location for the road.

From station 799 to station 822, distance 4 miles, the line runs on the side slopes of hills, varying from 10° to 30°. The grade will not exceed 6 feet per mile, and the section presents no difficulty whatever. The mouths of 17 ravines are crossed, the largest being 150 feet wide and 10 feet deep, the smallest 30 feet wide and 8 feet deep. No rock cutting in this section.

From station 822 to station 825, distance 5½ miles, the line runs partly at the foot of basaltic cliffs and partly at the foot of steep earth slopes crowned with basaltic rock. There is a regular beach of rock boulders along the whole of this distance, and the level line was run on this below high-water mark. An embankment with retaining wall can be easily made at the foot of these cliffs, the total amount of which would be 1¼ mile and 15 feet high, besides which there would be about 600 feet of rock cutting 5 feet deep. There are four ravines to be passed requiring embankments of from 100 to 400 feet in length and 12 feet high. The rest will be side-hill excavation.

From station 825 to station 836, distance 4½ miles, the line for 3 miles runs along gentle side slopes; for 1½ mile the side hills are at an angle of 45°, and covered in one or two places with loose rock, where retaining walls are necessary. The grades can be thrown above or below, as may be found best. There are three large ravines requiring embankments of from 100 to 150 feet and 10 feet deep. There will be about 200 yards retaining wall.

From station 833 to station 836, distance 1¾ mile, the line, as shown on the profile, was run directly across the sand-hills to the mouth of the Pelouse river, but would not be the selected line of location. Should the line go up the Pelouse to gain the table land it would be thrown on the side hill above; or should it pass the mouth of that stream it would keep below on the river bank, where it would run on gentle side slopes. The grades and amount of work would, of course, be widely different on the two lines. Neither, however, would present any difficulty.

Bench-marks were established at the "Monumental Rocks," at old Fort Walla-Walla, near the mouth of that stream, and in two places, one near and the other at the mouth of the Pelouse. All were cut in the solid rock, and the height above the base line on the last one.

It will be seen by the above report that the survey made by me divides itself naturally into three sections. The 1st, from the Umatilla (station 325) to the Walla-Walla river; 2d, from the Walla-Walla to the mouth of Snake river; 3d, from the mouth of Snake river to the mouth of the Pelouse.

In the 1st section, distance 22¾ miles, there are 6½ miles of difficult work, rock cutting generally, and requiring a special examination to determine the best line of location. The rest is either side-hill cutting or embankment in sandy earth, and presents no difficulty whatever. There will be 550 feet of bridging.

In the 2d section, distance 10¼ miles, there is no difficulty whatever. The line will pass over a sand flat, the excavation and embankment will not exceed 10 feet in depth at any point where there is no rock, and no structures required except a draw-bridge of 1,500 feet across Snake river.

The 3d section, distance 59¼ miles, presents much greater difficulties than the others, in consequence of the basaltic cliffs and broken ledges of rock which occur at the several points mentioned in the above report. None, however, involve greater difficulties than have already been overcome on railroads in the eastern States, and all of these places can be passed without tunnels and with easy grades. There will be 15½ miles of the above distance which will be nearly all rock cutting along these cliffs, and which will require a particular instrumental examination to determine the best grade with which to approach them, and the best location to pass with the least work. Of the remaining distance 39 miles will be in side-hill excavation with occasional rock cutting. The location of the line will be very easy; and the character of the ground will admit of very light grades, the highest not exceeding 30 feet per mile. Embankment will be chiefly across ravines, of which there is a large number, and perhaps three miles across flats, with an average depth of 15 feet, depending on the location of the road.

The amount of curvature will be very great in consequence of the tortuous course of the river. The least radius will be a thousand feet; and very flat curves can be employed with great effect in passing the basaltic cliffs to diminish the amount of work.

Basalt was the only rock found on the Columbia and Snake rivers in the limits of the survey. It is full of seams, much of it columnar basalt, and will, I think, be easy to blast. It is excellent for rubble-stone masonry and ballasting, and can be obtained in the greatest profusion, as well as the finest quality of sand. There are remarkably few springs above the level of the river; and the soil is generally sandy, but has abundance of excellent grass both on the river and the hills.

With the exception of a few willows of small size, there is no timber on the Columbia and Snake rivers up to the mouth of the Pelouse. Wood for railroad purposes can be procured, as is well known, from the Upper Columbia, the Yakima, and the Clearwater, and rafted down; and it may not be irrelevant to remark that the location of the line directly on the banks of two large and navigable streams will afford unequalled facilities for the transportation of the men, their tools, provisions, and the iron for the road.

The right bank of Snake river was the one surveyed, as it was reported to offer the least obstacles to a railroad location. The left bank seemed to be more difficult, looking at it across the river; but it is almost needless to say that no final location could be properly made until both sides had been instrumentally examined.

I am, sir, with much respect, your obedient servant,

W. W. DE LACY, *Civil Engineer.*

Lieutenant JOHN MULLAN,

U. S. Army, Commanding Military Wagon Road Expedition from Walla-Walla to Fort Benton.

I.

FORT OWEN,
Bitter Root Valley, W. T., January 8, 1860.

DEAR SIR: Agreeably to your instructions of November 5, 1859, to make a reconnoissance eastward from the St. Regis Borgia river to Fort Benton, in order to ascertain in detail the character of the line for the proper location of the military wagon road, and in order to arrange the working parties for spring's operations, and at the same time collect data regarding the climatology of the main range of the Rocky mountains in midwinter, I left your camp on Wolf's prairie, on the St. Regis Borgia river, with Major John Owen, Indian agent to the Flathead nation, November 7, 1859, and reached Fort Owen, after seven days' march, on the 13th, estimating the distance at 132.5 miles. The weather becoming very cold, and the snow falling continually, we had in that short journey many little difficulties to overcome, which delayed us considerably, and thereby impressing me with the idea that a reconnoissance under such circumstances would bring but poor results; I therefore addressed you November 14th a note, in which I informed you of the disadvantages under which I had to undertake the work, mentioning at the same time that I would remain at Fort Owen until I received further instruction from you; A sudden change in the weather induced me, however, to determine to start, but the difficulty to find a person who knew the wagon road trail delayed me until the 22d of November, at which date I left Fort Owen, accompanied by C. E. Irvine, esq., two laboring men with six riding and four pack animals. I made that day twenty miles against a strong cold north wind, and camped on the Bitter Root river November 23; two of the riding animals, belonging to Mr. Irvine and myself, had strayed during the night. I made a late start, 9 a. m. Mr. Irvine turned the two missing animals, in case they should be found, over to Mr. Jacobs, to go down to your camp; at 11 a. m. I crossed the upper Hell-Gate crossing, at which place the river is divided by an island. The stream was heavily floating with ice, and the crossing rendered difficult by a strong ice cover extending from both shores some twenty feet into the river. The day was cold, and a strong wind blew from and through the cañon. In three-fourths of a mile we came to Rattlesnake creek, which is one-fourth of a mile from the mouth of the cañon, and found much difficulty in crossing it on account of ice; from here the ground rises very gradually for one and one-half mile, at the end of which a steep descent of some sixty feet to the lower plateau would require some work; if the road through it kept near the river, which will occasion a little detour, this place can be avoided. The ground being gravelly and loose, any earth-work in the lower portion of the defile would be easy, and quickly accomplished. We passed over said lower plateau for one-half mile, and came then to a steep ascent, (sixty or eighty feet,) which cannot be avoided. There is a kind of ridge running towards the river, with an almost level top of forty yards in width at the foot of its westerly side, which has a steep slope of sixty feet. The road crosses a small creek; the ascent on the other side is about twenty-five feet, then three-fourths of a mile level ground, then crosses White Thorn creek with almost undistinguishable banks, then one mile level ground about thirty feet above the surface of the water, thence to the crossing of the Big Blackfoot river, twenty yards above its mouth. No difficulty for wagons to get in and out of the river. Depth of water, two feet in the channel. Between this river and the first Hell-Gate crossing, which is five and one-fourth miles distant, the road leads through open timber and level ground. Two creeks have to be crossed, the second one with miry bottom; after making the river crossing, which is a good one, the character of the country is in every respect favorable for wagons. The ground is level, firm, and is covered with open timber for two miles. I then reached the second crossing; the river here forms many arms, which overflow at times the enclosed islands, and renders them in places miry. Then for three-fourths of a mile through open timber to a little creek one-fourth mile from there, reached a prairie one and three-fourth mile in length and one-half mile in width, at the end of which I crossed a little creek and went to camp, made twenty-three miles, snow on the ground about two inches deep; great many icy places on the road to-day, where the snow has melted and frozen again. A little snow falls towards evening.

November 24.—It snowed during the night, and we started at 9 a. m. in quite a heavy snow-storm. After three miles travelling through open timber, came to a prairie three miles long, at the end of which we crossed the river (third crossing) which forms an island, and one and one-tenth mile further, crossed Rocky creek at a point where it also forms an island. This creek is rapid, large, and two feet deep, with big boulders in its bed; then three and one-half miles open timber, then three-fourths of a mile over a piece of ground which cannot be called either prairie or timber land, the trees standing at such a distance from each other that the road can be laid straight over it, thence to the fourth crossing; thence three-fourths of a mile through open timber, with a good deal of underbrush, thence across a prairie three and one half miles long to a creek which runs along Beaver-tail Point. This is a tolerably high and steep hill, which will require some work to make it practicable for heavy wagons; the hill forms a single spur of one-half mile in width, terminating at the river in a perpendicular bluff. The present wagon road makes here two crossings of the river to avoid this place, (fifth and sixth crossings.) Then one and one-fourth mile over prairie bottom to the river crossing (fifth) by the trail and (seventh) for the present wagon road. This crossing might be avoided by keeping the road on the right bank, but it would require some earth-work and the bridging of a little creek which runs in a steep ravine. This creek has been named Gold creek, as Colonel Lander is said to have found gold specimens in it. One-fourth mile from the fifth crossing (by trail) a bad slough has to be crossed; then one and one-fourth mile over prairie bottom to the sixth crossing, (eighth by present wagon road;) then one-fourth mile over prairie to Little creek; then one-fourth mile over prairie to the seventh crossing, (ninth by present wagon road;) then one-fifth mile through open timber to the eighth crossing, (tenth by present wagon road;) thence the trail takes over a side hill, about 250 yards from the last crossing.

The wagon road keeps in the bottom, and makes the eleventh crossing a little below, where the low water trail crosses, the wagon road would make the twelfth crossing. After going for one half mile through open pine timber, and then go over a fine prairie one-fourth mile in length to the thirteenth crossing. Instead of following the low water trail, which forks here, one part making the river crossing, the other winding around the rocky bluff which springs out into the river, I took to the high water trail, which led me high up on the hill side. This trail turns, after striking the above mentioned prairie, over a little swelling of the bottom, to the river crossing; thence one-half mile through the bottom, which is cut up by a miry slough, towards the hill bluff, which has a steep slope. At this point the wagon road will keep a little to the south, or right of the trail, to gain the saddle of the projecting ridge. This place will require some earth-work that cannot be avoided, and has offered a great deal of difficulty to all the wagons which have come yet from Fort Benton. Reaching the opposite foot of this hill, which is one-fourth mile from base to base, the road gains in three-fourths mile, going over prairie ground, a slough which is said to be very bad in summer time. Then strikes over a prairie three miles long, called "Cold prairie," to a ridge running in the same direction as the first one. This spur, which is not so high as the previous one, can be avoided by the wagon road by keeping close to its foot around the point. This, though, will involve a detour of three-fourths mile, and I think, with little work, the ascent of it can be made by taking advantage of a little ravine which runs up the hill side close to the trail. After reaching the top of the hill a very gradual descent of one-half mile over prairie ground brings the trail to a little creek; then three-fourths mile over prairie to an old river bed, or to a lower river bottom; then one-half mile along this bottom to camp; made 27 miles. Snow two and one-half inches deep. It snows a little towards evening.

November 25.—Snowing all night, and still continued this morning. Started at 10 a. m., our horses having strayed. Cold day. Went over the prairie bottom for three-quarters of a mile; then the trail takes to the hills, and crosses in one-half mile a small creek. The wagon road can avoid this hill by keeping around it, and striking the trail again a out one-half mile beyond the creek crossing. Then for one and three-quarter mile over rolling prairie ground to two little creeks, which run close to each other, separated only by a small elevation; both of these creeks have miry bottoms; the Hell-Gate river being separated by a bare ridge from the road, and is completely hidden from view. Then for one-half mile over rolling prairie to another creek, taking, immediately after crossing it, to a considerable prairie ridge coming from the south, and forming a connexion with the hill range, which separates the road from the Hell-Gate bottom; some grading will be required here, but it is light work. Following that ridge for two and a half miles in an easterly direction, with its ups and downs—all of which will require some side-hill cutting—we reached a small creek; then over a small gently sloping hill to a small spring branch—from creek to creek about eighty yards; then take to the opposi e hill, where also some grading will be found necessary, and in one-fifth mile reach a second spring.

Ascending the next hill, requiring also side-hill cutting, the Flint Creek valley lays before us, whence I commenced the gradual descent, and reached it after four and three-quarter miles' travelling. Flint creek has little wood on it, but still good camping places may be found above the crossing place. The creek is two and a half feet deep, but the banks not high. After crossing Flint creek, in one-tenth mile we crossed a small branch of it; then for one-quarter mile through the valley bottom; thence three-quarters of a mile over rolling prairie hills, forming the river bluff of Hell-Gate river to the valley itself. The descent from these prairie bluffs, at the place where the trail strikes it, is rather steep, and requires cutting; but I am confident that by a little examination a better place could be found. Following up the Hell-Gate bottom for one and a half miles I camped on a creek; after crossing it, the trail bends about according to the shape of the creek, which makes an elbow towards the north. In two and three-quarter miles cross a second creek, and then follow up the valley for three and one-fourth miles, winding part of the way around the swampy bottom of the river. This portion of the valley is cut by a great many ditches, all of which are filled up with drifted snow often to the depth of four feet; the snow at a level in the valley is five inches deep. The trail is unbroken, and our animals have had hard work to keep up a good marching trot. After crossing a hill spur of three-quarters of a mile in width, with gentle slopes on both sides, I took up the bottom again for two and a quarter miles, and then went to camp; made twenty-two and a half miles. The day has been clear but cold. The prairie bluffs are almost bare of snow, which has been deposited by the wind in the valley. I am confident that the wagon road can be located in such a manner as to run almost straight, as the valley, which is one and three-quarters of a mile wide, and the river bed, though winding much, keeps well to its centre, without touching the bluffs.

November 26.—Started at 8½ a. m. Two mules have strayed off. Cold morning; snowed heavily. Travelled one mile through a bottom; then crossed "Humbug creek," which has a fine valley running up towards the pine-clad mountains; crossing this valley, reached, after three-eighths of a mile, the fourteenth crossing of the Hell-Gate river; this crossing we made in the ice. All the former crossings and those of the Big Blackfoot are rocky.

Flint and Humbug creeks have offered us more or less difficulty, as the shores were frozen on both sides, requiring a leap from the ice into the water and out again on the icy shore, which always endangered our animals, the ice being bare of snow. Hell-Gate river here is a bold, rapid stream, from eighty to one hundred yards wide, with a channel of two feet in depth. After making the crossing I travelled for two hundred yards over a prairie bottom; then crossed a small timbered point; then took to a little side hill, which will require some side-hill cutting. The present wagon road makes at this point two crossings, being not more than one hundred and twenty yards apart. The trail, after leaving this side hill, will require more work, being rocky and longer than the first one mentioned. At this place the wagon road makes two more crossings of the river; then one-quarter of a mile over prairie, up the bottom, to a place where the road winds for one hundred and eighty yards around the side hills, to avoid the crossing of a miry slough formed by a spring; around those hills, which are a little sideling, very little work will make this a good road. Then for one and a quarter mile over level prairie, up a bottom; then cross a creek; then follow up a prairie bottom for one-third mile; then cross a coulée, which wagons can avoid by keeping to the left of the trail; then go up the bottom for three-quarters of a mile; then cross a little hill spur, which may possibly require grading; then for two hundred yards over prairie bottom to a creek; then up a level prairie bottom for two and a quarter miles; then the river takes a short turn to the south, and there also the trail forks—one passing up Deer Lodge valley. The wagon road here leaves the river, and crossing the valley takes to a gently sloping side hill, and after one and a quarter mile strikes the Little Blackfoot river about one and a half mile above its junction with the Deer Lodge creek. From the top of said hill the junction of these waters can be seen, and also the lower portion of the big valley. None or very little work will be required to bring heavy wagons up to the top of the hill without doubling teams; but it may be that some winding about in the descent would be necessary, as some deep hollows cut up the easterly slope. From the foot of the hill one-half mile through the bottom of the Little Blackfoot valley we camped, having made eight and a half miles.

Mr. John Grant has settled near the junction of the two branches of the Hell-Gate river, and built two log-houses. It snowed since 4 a. m., and for some time pretty heavily; at 11 a. m. it cleared off. The snow here is not so deep as below; it measures, on a level, three inches.

November 27.—Very cold morning. Started at 9.45 a. m. Had much delay in finding the animals; one mile up a prairie bottom; then one-eighth of a mile over a side hill, which will require cutting; then one-eighth of a mile up a bottom, partly through brushwood; then winding for a quarter of a mile around the river bluffs, where work will be necessary; but in keeping the road along the right bank six crossings of the river would be avoided. Gaining the hill-slope the road strikes in half a mile a creek, and crosses in half a mile further a second one, the intermediate space being slightly undulating ground. After ascending a gently sloping hill, I followed its rolling surface for two and a quarter miles and began then the very steep descent. Here a careful examination of the ground would have to be made to find the best location for the wagon road. This descent of about 350 feet offers one of the greatest difficulties on the whole route. Of the two ravines to the right and left of the trail, one or the other may be adapted to the purpose, if not, considerable work, similar to the descent to the St. Joseph's river, will be found necessary. After reaching the foot of the hill the road runs up the river bottom for three-quarters of a mile and crosses at that distance the Little Blackfoot at a point where it forms a timbered island a quarter of a mile in width; both crossings are good and with gravelly bottoms; depth of water about one foot and three-quarters and width of from 15 to 20 feet. The opposite hill, which the road has to cross, is called the mountain, and said to be more formidable than the divide of the Rocky mountains. From the foot to the top of the hill the distance is one eighth of a mile, and I am confident that by two turns, which will involve some side-hill cutting, a good and easy grade can be attained. The present road does not ascend the hill at the place where the trail is located, but keeps for a mile further up the valley, which forms here a kind of cañon, and then ascends the hill by taking advantage of a coulée, which runs up to the top of the mountain, with little work. It is said that a good road can be made there, but it will make four river crossings necessary.

From the top of the hill, for three-quarters of a mile, over gently rolling ridges, to a timbered point of one-tenth of a mile in width; a quarter of a mile further, over ground of the same character, I gained the topmost ridge; then commenced the very gradual descent, over an undulating surface, towards the Little Blackfoot River valley, which I gained in two and a quarter miles. At the foot of the last slope crossed Irvine's creek, which has miry shores. At this point the wagon road could, after crossing, follow down the bottom of that creek, then cross the Little Blackfoot river and strike over an apparently easily graded hill towards the north fork of the Little Blackfoot river. This road, though it is said to be a bad one in its present condition, is worth examining, as it would save four or five miles in distance.

The present wagon road takes in a quarter of a mile after crossing the bottom of Irvine's creek to a gradually ascending hill spur, the surface being covered with stones, which it follows for one mile and a half, requiring no work at all to bring wagons over; then cross Buffalo-head creek; then following up the level bottom of the Little Blackfoot for one mile and a half, reach a slough, and in one mile further a second one, which must be very bad in spring and summer. In three-quarters of a mile, still following up the valley, I crossed the Little Blackfoot river at Belknap's camp, after passing a plateau which is elevated about six feet over the level of the valley. Ascending the almost level valley on the right bank of the river, I reached, in three and a half miles, a kind of terrace or high bottom, 250 feet in length and terminated by the prairie bluff, coming in close to the river and forcing the present wagon road to make two crossings; with little work the road, though, can be kept on the right bank. Passing this side hill the ground is level for a quarter of a mile, and then another side hill closes in, which will require some work for 85 to 90 yards. At this point two more crossings are made by the wagon road. In order to avoid a small rocky ridge which springs out into the valley the road turns to the northeast, and passing the saddle of the ridge, crowned by a few pine trees, strikes, in three-eighths of a mile from the turning point, the north fork of the Little Blackfoot river. The forking point is about half a mile below; here, again, the present wagon road crosses the river twice to avoid two side hills, which in the space of half a mile runs in towards the river. I think, though, that the little work that would be necessary to keep the road on the right bank would not justify the crossing of the river, which seems to have miry shores. Came to camp after having made 19 miles. The day has been very cold, and towards evening a chilling northeast wind makes the cold still more severe. Snow varies in depth from two to four inches.

November 28.—Started at 9.20 a. m. Cold, but clear, morning. Up the valley for 200 yards, then turn a sharp side hill, and immediately afterwards ascend a very sideling hill. From the place where the sharp point comes down to the easterly foot of the second hill, three-quarters of a mile distance; then up the valley, which becomes very narrow for one and three quarters of a mile; the trail winding for the last three-quarters of a mile around the side hill. The wagon road, though, as the ground appears to me now, can keep lower down, thereby avoiding work. At the end of that distance, a bad rocky point has to be passed, and I think the wagon road could cross here, with advantage to its location, and keep for three-quarters of a mile on its left bank. First turning, at a distance of one eighth of a mile, a gently sloping hill spur, and the crossing, a half mile further up, a small tributary of the Little Blackfoot, where the road could recross to the right bank, is a fine prairie, at a place where the valley widens considerably. The trail, which runs along the right bank, crosses a half mile from the last rocky hill, spur, or creek, which runs in a valley leading north to a spot where the Little Blackfoot valley is strongly bending to the east. Should the road be kept on the right bank of the river, the work is loose rock. Around the point where I propose to make the crossing would extend from 50 to 60 feet; from there, for two and three-quarters miles, no earth work would be necessary, only a corduroy of 35 feet over a wet place formed by a spring coming from a hill spur. Two hundred and sixty yards from there, another wet place, originating in the same manner, can be avoided by keeping closer to the creek, where the ground appears to be dry. I think it necessary that this portion of the road should be carefully examined, because the valley is said to be miry in the extreme during spring and summer, and probably it will be inevitable to cut the road here high up on the side hill. From the last-mentioned wet place, a half mile up the valley bottom, requires no work at all; here the road must cross the creek and keep on its left bank all the way up, as a long hill spur, which would require a quarter of a mile earth work, runs close on to the creek, one mile and a quarter from where I propose the wagon road to run. The trail crosses the creek also, taking to the valley toward the divide, leaving the Little Blackfoot valley, which extends to the north, up the hills to the left. Following the eastern valley for an eighth of a mile, I came to a ravine of an eighth of a mile in length, which, though narrow, will offer no difficulty to the location of the road, crossing the little water-run which empties at the last-mentioned crossing of the Little Blackfoot. I went for three quarters of a mile over a level plain, which is situated just at the foot of the divide; then for three-eighths of a mile to the top of the divide, where I found the snow from four to five inches deep. On the westerly slope, in some places, the snow has a depth of from six to eight inches; the easterly slope was covered with only from two to three inches. A strong northeast wind struck us as soon as we gained the above-mentioned plain, lulling only towards sunset.

The natural surface of the ground, though open timbered, will probably require no work, though in places the grade is somewhat steep. At the foot of the divide, a quarter of a mile from the top, I struck the headquarters of Big Prickly Pear creek; one mile and three-quarters further, a little branch of running water, coming from the north, crosses the road. Just before reaching this creek a little side-hill cutting will have to be done, and also on the other side of it are points where the hills come close to the creek; the valley widens considerably and affords good camping places. On the east side of Stormy creek the ground is level for three-eighths of a mile; then a side-hill cut of 20 yards will bring the road again in an almost level plain for three-quarters of a mile; at the end of which the wagon road leaves the Indian trail, which leads down the Big Prickly Pear, and takes to the west over a small side hill covered with pedrigal rock, and in a half mile crosses a dry water-run. An eighth of a mile further a second one is crossed; taking then a half mile over a gently sloping hill, covered with pedrigal rock of large size, a thin water-run is crossed; ascending from there a gently sloping hill, covered with a few red cedars. I gained, in a half mile from the last dry water run, the narrow valley of Fir creek, a tributary to the Big Prickly Pear creek; a few steep places in the descent will offer no difficulty for wagons. The ascent on the other side of Fir creek is very steep, but a ravine, running to the right of the trail, will facilitate the location of the wagon road; an eighth of a mile is the distance from the creek to the top of the hill, which slopes very gradually. On the opposite side, for a quarter of a mile, towards a dry pond, which extends to the next hill, of easy ascent and descent. The latter one terminated by a dry water-run; distance from the spot where I struck the dry pond to the dry water-run is three-quarters of a mile.

The next prairie hill is tolerably high and of a grade so steep that doubling teams will be necessary for heavily laden wagons. At this ascent, if possible, the road should be located at another point, as the present trail is cut up by many narrow but deep coulées. Ascending a hill one quarter of a mile long, from the top of which is a fine view of the Upper Missouri valley, the gate of the mountains and of a small portion of the Rocky mountains, Heart mountain, can be distinctly seen, and also the direction of the wagon road for ten or eleven miles, to a point where it crosses Soft Bed creek. After crossing a dry water-run we passed over some rolling ground and descended to Silver creek, one of the tributaries to the Big Prickly Pear creek; thence to camp. Distance from top of last hill to camp four miles and a half. Made, to-day, nineteen miles and three-quarters. Our animals are getting weak; the want of sufficient nourishment commences to tell on them. The day, which commenced cold and cloudy, became clear as soon as the wind sprang up, and towards evening it became quite mild.

November 29.—Started at 9¾ a. m. Our horses had wandered far off during the night in search of grass. Mild, cloudy morning. For five miles and three-quarters passed over undulating, firm prairie ground, to a small creek with miry shores, bordered by small willows, whence its name of Willow creek; thence two miles and a quarter over very gently downward-sloping prairie, part of which is covered with brushwood, to Soft Bed creek, a branch of Small Prickly Pear creek. After crossing it, I went up its valley in an easterly direction for two miles, at the end of which a little side branch will have to be crossed. For the last half mile the side hills close in so narrowly to the creek bed that in some places hill-cutting will be found necessary, but the distances where work is required are short and the ground not difficult. For one-fourth of a mile further the character of the ground remains the same as described; then, by turning a little more to the north the road gains the hill-side, and gradually ascending reaches in two miles the top of a little divide. The road before coming to the last ascent, the trail to its right taking advantage of a small ravine, which facilitates the ascent considerably. The valley on the other side is quite narrow, and remains so for half a mile, until a point is reached where the creek will have to be crossed and immediately recrossed. A little work now and then in this distance will be necessary at points where the side hills are too sloping. At this place the valley widens considerably, and a good camping place for a large train can be had. I suppose, though, that the valley must be wet, as the trail keeps on the westerly side hills and recrosses the creek only three-quarters of a mile below where the valley bends to the north. In fifty yards the trail recrosses again, and remains on the west side of the creek all the way down to the Little Prickly Pear creek. In two miles and a half more it passes along in an easy slope, which is occasionally cut up by small ravines, but offering no difficulties to the wagon road. Little Prickly Pear creek, which runs at the place where I struck it in a large valley, is twelve feet wide and one foot and a half deep. The banks of the creek in this plain are lined with willows, but higher up and lower down large trees will be found. The creek, coming from the Rocky mountain chain, cuts through a narrow cañon to join the broad bottom, and forces its way after three miles through a rocky mountain range, forming a narrow gorge. After crossing the creek, we proceeded in a northerly direction towards Medicine rock. The day had become very warm, and we had hoped to have been able to reach Mullan creek before dark, but after riding for one mile a sudden northwesterly gale forced us to seek shelter in the willows of the creek just crossed. In less than half a minute our horses were covered with white frost, and the air was filled with an icy, foggy substance, which after some time gave way to a heavy snowfall. Recrossing Little Prickly Pear creek, we found a tolerably good camp in the willows. We made, to-day, fourteen miles headway, but travelled sixteen miles.

November 30.—Remained in camp to-day. The gale has not abated in the least, and the snow still falls in abundance.

December 1.—The night has been cold and stormy. At 9 a. m. the sun broke through the clouds, and I started at 11 a. m. It was very cold, and the snow was five inches deep; and the road leading to Medicine rock strikes in three-fourths of a mile a little creek, and following it up for half a mile strikes the foot of the hill range. Here the trail battles a very steep ascent, which the wagon road can make more gradually in keeping to a ravine running to the left of the trail. Reaching the top of the first hill, a good road with easy grades leads towards Medicine rock, and from there descends to a small creek called Medicine Rock creek. The descent will require some side-hill work in places where the slope is too steep. From the foot of the hill to the creek is two miles. The next three miles of the road, which brought us to Mullan's creek, will require work in different places, as the trail keeps along the side hills. Two hollows, the first one pretty deep, will have to be passed, and a few trees to be cut, to bring wagons through with four or five yoke of oxen. The ascent on the other side of the creek is steep, but short, and then the character of the ground for two miles further remains the same as described. At the end of that distance the road leads towards Prickly Pear Creek valley, thence over a bare hill, the descent of which will offer no difficulty except in one point, which, for the present, is made passable by a couple of logs stretched from side to side of a low rocky precipice, to give in that narrow passage the required breadth to the road. Reaching the valley, I crossed a creek coming from the west half a mile distant from the bending point of the trail on top of the hill, then half a mile through the bottom of the valley, keeping close to the hill-side, through which, in connexion with the vegetation of the valley matters, we believe that it is of a miry character. Taking again to the side hills, I reached in two miles the head of what is called Prickly Pear cañon, and went to camp. The last portion of the road will also require some side hill work, which in some places may prove more difficult, as the ground is rocky. Made, to-day, 11⅝ miles.

December 2.—The wind blew a perfect gale last night. The morning is very cold, and our animals appear much reduced. The snow in the valley is four inches deep and covered with a hard crust, which makes pawing for the grass by the animals difficult. At the head of the cañon through which the road passes a small creek, coming from the north, empties into Prickly Pear creek. By information I learned that by following up this creek a good road can be obtained, avoiding thereby all the crossings which are inevitable by following down the Little Prickly Pear. In taking this direction a little detour might be made, but it is said that no obstacle of any moment would be in the way. Started at 9.22 a. m. At the entrance of the cañon, where the first crossing will have to be made, after having crossed the creek alluded to at its mouth, we found that the usually travelled road here has been made impracticable by beavers. Finding a good place to cross, we followed down the cañon for eight and a half miles, making seven crossings, all of which were greatly obstructed by ice. The intermediate ground is partly covered with pine timber and partly with brushwood of different heights. After the seventh crossing the cañon becomes wider and forms a valley, affording ample room for camping places, with good grass on the low side hills for the animals. In half a mile from the seventh crossing I reached the eighth, which is situated at the foot of the northerly hill bluff. All the crossings are shallow with firm gravelly bottoms, and the banks offer no difficulty. After ascending the low bluff, which will be easily rendered practicable for wagons, the road leads over prairie ridges with easy slopes, to a small creek two and a half miles distant, which empties into the Missouri river a few hundred yards below the Little Prickly Pear creek. From the crossing point the road follows the creek for three miles, for which distance no work will be necessary, except at one point, where the side hills close in narrowly. At the end of this distance a low but steep ridge is ascended by the trail. Before reaching this point the wagon road must cross the creek and follow it for three-eighths of a mile higher up, then recross the creek and surmount the ridge, the second hill spur, which is the largest, but is of easy grade, offering no difficulty in its slope. The descent on the other side, which terminates at a dry creek three-fourths of a mile distant from the one I just left, is in some places steep, but not sideling. The next ridge, measuring from base to base three fourths of a mile, is not so high as the previous one, nor is it as steep. After crossing a dry branch I reached, in two and three-fourths miles, travelling over undulating ground which would require no work, another dry water-run. At this point the above-mentioned road, which leaves the one here described at the head of the cañon, would again fall in with the present wagon road. In three miles further I struck the Dearborn River valley; the ground is undulating prairie for this last distance, and no characteristic can be mentioned regarding it. Half a mile from the Dearborn valley a small creek will have to be crossed. A high and well-timbered ridge to the eastward indicates the Missouri River valley, and a curiously formed hill range. To the north of the Dearborn are good landmarks for the traveller who cannot form his course by the uniformly undulating plateau. The road will have to keep up the Dearborn River valley for one-half mile, to come opposite to the crossing point. The valley itself is richly timbered, but cut up by a great many sloughs and ditches, which run in every direction. Dearborn river is a shallow stream of from fifty to sixty feet in width, with a gravelly bottom and swift current. The wind has been blowing hard and cold all day from the southwest, and is still increasing in force; made seventeen miles; snow five inches deep.

December 3.—It has been a windy night; our tent was blown down, and I was fearful that some of the shallow-rooted bottom trees would be blown on top of us. This morning it is quite mild. Started at 1¾ a. m. to make Sun river in good time. The ground is generally undulating from the north bank of the valley to the crossing of Beaver creek, which is five miles distant. The road first follows up a dry creek, crossing and recrossing it several times, and then leads over a wave-formed plain, which now and then is cut up by long, dry hollows. After crossing Beaver creek the trail commences the ascent of the divide of the Dearborn and Sun river waters; and, if I understand aright, the present wagon road keeps to the left, making a detour of four and five-tenths miles, avoiding thereby the steepest portion of this rocky divide. Just as we commenced the ascent an icy northwest gale sprang up, driving before it a storm of small crystals, which made it almost impossible to keep the eyes open. The wind continually increasing in force, we became doubtful whether our animals would be able to carry us to Sun river, and were obliged, after having made two and a half miles more, and nearly gained the summit of the divide, to turn back and seek refuge in a small ravine, where we found just wood enough to build a fire and stow away our baggage. Made seven and five-tenths of a mile. The storm continues with fearful severity; it is extremely cold, and the snow falls heavily. We have an idea that we may be caught here by a long lasting snow storm, which is anything but pleasant, as wood is scarce, provisions nearly exhausted, and our animals are in a distressed state.

December 4.—The weather has not moderated in the least, and I judge the thermometer must be at about 40° below zero. It still snows, but not so heavily as yesterday. We are undecided whether to push on to Sun river or go back to the Dearborn. To retrace our steps seemed the most sensible in our present condition, as we would have the wind in our backs, and in a short ride could reach shelter and food in a large Pend d'Oreille camp, which we left at the Dearborn river; but then the snow might fall to such a depth that we would be cut off from Fort Benton. At 10 a. m. we decided to try to reach Sun River farm, and at 11 a. m. started for that point. It is terrible weather, and we can distinguish nothing at a distance of one hundred yards. After a ride of three and a half hours we reached Sun River farm, which was twenty-three and one-tenth miles distant from our last night's camp. We were all more or less frost-bitten, and have suffered extremely. Colonel Vaughan, Indian agent to the Blackfoot nation, received us very kindly, and with his well-known hospitality offered us the accommodations of the agency for any length of time that we might wish. The weather continuing very cold, and our animals requiring some rest after our last march, we remained at the farm during the 5th and 6th, and left on the 7th of December. The cold being too severe to take notes on our march, I did so (for this portion of the road) on my return; but will give the description of it here, in order to keep the same travelling direction in mentioning the different objects of the road.

From the crossing of Beaver creek to the top of the divide, which forms a narrow rocky gap, the distance is three-quarters of a mile. It is not one gradual slope, but the ascent is formed by a succession of hills of different grades and heights, which will occasion some work in a few places if the road should be located in that direction. Hitherto it has been thought a matter of imposibility to follow with vehicles of any description the Indian trail, but in coming back from Fort Benton I brought over it a two-wheel cart, and found but little difficulty. The descent on the other side is gradual, and is practicable for wagons in its present state. Bird-tail Rock creek, a little stream which heads near the divide, is followed down by the trail for one and a quarter mile, and then at a point where it turns short to the west the trail takes for two and half miles over prairie hills, until it strikes the creek again, and also the present wagon road, which gains the creek about one and a half mile below the above-mentioned turning point, and follows its valley downwards, crossing and recrossing the creek, which is made necessary by the closing in of the small side hills, the trail comes in three miles opposite to Bird-tail creek, a prominent peak of the mountain chain running between the trail and the Missouri river. Following for three miles more the valley of the creek, I reached a low rocky bluff, through which the creek had forced its way. Here a little side-hill work will be necessary, and also on some places above. In three and a quarter miles over an almost level plain the road crosses Crown-butte creek, about two miles below the mouth of Bird-tail Rock creek, and gains in one and three-quarter mile a small rocky ridge, which is easy in its ascent; the descent is sideling, and encumbered by big boulders, which will have to be removed. The Big Knee, Crown, and Square buttes, which have been in view for some miles back, are now fully exposed to view. A kind of saddle ridge, which connects the Big Knee and Crown butte, with a gentle slope, will have to be ascended. The road leaving those two prominent points almost at an equal distance to its right and left, takes then to the descent, which is very gradual. The direction which I followed brought me opposite to the agency; but I think the wagon road must keep closer to the Big Knee, and strike Sun river about three miles below the farm, to gain a point where the crossing towards Teton river can be accomplished without distressing the teams. Sun river, which I crossed at a point where it forms an island, was frozen over. A luxuriant growth of cotton-wood lines both sides of the stream at this place, and extends far upwards. Distance between the last-mentioned little ridge and Sun River farm is eighteen miles. By information, I learned at Sun river that the distance to Fort Benton by shortest route was estimated at forty-five miles. At nineteen miles from the farm the road strikes a shallow pond, and seven miles further is a small spring. Both of those places, though, are destitute of timber. In summer the pond often dries entirely up, and, at all events, the water becomes unfit for drinking and cooking purposes; besides that, in both places it is strongly mixed with alkali. In the strength of those statements, I resolved to try and find a wagon road leading to camping places where wood and good water could be found, and concluded to strike the Missouri river somewhere below the Great Falls. The interpreter of Mr. Dawson being at the time at the farm, I accepted his offer to guide us to the fort.

December 7.—Started at 10.15 a. m.; mild weather. Followed down the Sun River valley for about 12 miles, and then struck across an insignificant prairie ridge towards the Missouri river. Before I left the valley, which we did at a point about three miles above the mouth of Sun river, I crossed a deep ravine, with running water, ten miles from the farm. After 18 miles we arrived at the edge of the prairie plateau, and had to accomplish the descent towards the Missouri river by following the winding of a coulée ridge, gaining the coulée bottom itself, which quickly enlarges to a narrow valley, with a few cotton-wood trees. We reached in one mile the river itself. We intended to follow down the Missouri for one and one-fourth miles further and camp on a small prairie opposite the mouth of Heighwood creek, but the river ice was blocked up over the trail, and we were forced to remain where we were, having made 31 miles.

The Missouri river is so poorly timbered between the mouth of Sun river and Fort Benton that if even a road could be found in that direction the greatest inconvenience for want of wood would ensue, even for small parties. During the night a furious gale forced us to secure our blankets and baggage with logs and rocks, and even then those precautions proved to be insufficient.

December 8.—Started at 9 a. m. Our guide advises us to head the coulées, and we therefore started back to gain a ridge about five miles to the northwest. The day is clear and warm. After a ride of 28 miles, during which we only crossed the Grane, and then the eight mile coulée, we arrived, after descending near Discovery butte into the Missouri River valley, at Fort Benton, and were most kindly received. There was scarcely any snow here, but the river is frozen solid, which is a fortunate circumstance, as we can drive our animals across, who would starve on this side.

The necessity to give rest to my animals, and some delay in my preparations for the return trip, induced me to remain seven days at Fort Benton, during which time I received from M. Clark, esq., of Fort Campbell, a small howitzer belonging to the government, which I intended to bring over the mountain, and to measure the road by its wheel.

December 15.—Mr. Dawson having fitted out a party to accompany me to the Bitter Root valley, I started at 12 a. m. towards the Teton river, *via* Sun River agency. Colonel Vaughan, who arrived at Fort Benton on the 13th, directed his returning wagon to keep also with us.

The odometer was fastened to the cannon wheel, which measured nine feet ten inches in circumference. After ascending the Missouri river bluff, which has an easy grade, we followed a well-beaten wagon trail towards Teton river, over almost level ground, and struck said river in four miles. Before descending into the valley the road forks, one keeping along its right bank on the bluff, the other descending into the valley and following it up. Our guide for this portion of the road having been kept back at Fort Benton, we took to the valley, which afterwards turned out to be the longest and most troublesome road of the two, on account of the furrow-like frozen surface of the river. In three and one-tenth miles we made five river crossings, which delayed us considerably, as we had to throw earth on the ice to give our animals a foothold. No other difficulty for wagons was experienced. Made to-day seven and seven-tenths miles, and camped on Teton river. About two inches of snow on the ground, but the grass has been mostly burnt off, and partly eaten off by Indian horses.

December 16.—Started at 11 a. m. Our animals had gone towards the Missouri river bottom, and it took us a long time to find them. We made six crossings to-day, accomplishing it by great labor. After following the valley for six miles we took to the north bluff, which offered an easy ascent. The top of the bluff forms a large plain, and the river, which makes a wide bend towards the south, disappears out of view. In the distance the Teton buttes can be seen, the trail taking a straight shoot, towards the westerly end of the river bend, heads all the coulées, offering thereby an almost level road. In five and one-half miles we struck the river again and descended its valley, and camped in one and one-half mile further on the bank of the Teton.

Baptiste Champaigne, our guide, who joined us this morning, informed me that one-fourth of a mile below, and opposite the spot where we struck the river valley again, the straight road from Fort Benton descended also to the valley, and that the distance to that place was estimated by Governor Stevens to be ten miles. Where that road strikes the valley the river will have to be crossed for the first time, gaining thus nine and two-tenth miles in distance. Wood is plentiful, the valley large, and covered with good grass. A rain storm passed over us towards evening, changing by nightfall into snow. The snow kept up all night, we having made to-day 13 miles.

December 17.—It is still snowing. We started at 11 a. m. To-day I attached the cannon to one of the wagons. The road follows up the valley for three miles, where no other obstacle is offered but a little creek running in a hollow, which can be made practicable in a few minutes' work. A second ditch at the place where the road leads towards the bluff the same amount of work will be needed. The ascent itself, formed by a gentle slope, requires a little side-hill cutting for a distance of twenty feet, to make the road practicable for heavy wagons. The road on top of the bluff requires no work at all, being almost as level as a table. In eight and a half miles I gained the descent to the river, which will require a little side-hill cutting for a distance of twenty-five feet. A half mile further I camped, having made twelve miles. The day has been very cold ; the snow kept falling heavily, and the wind blowing strong from the north. Snow four inches deep.

December 18.—Started at 10 a. m. Clear and cold morning. At the start our wagons were obliged to make the river crossings to avoid a side hill of thirty feet in length, requiring but little work to make it passable for wagons. In three-quarters of a mile from camp I came to the last Teton crossing, which was frozen over like the rest. The river is about twenty yards wide, and appears to have a gravelly bottom. The banks at the crossing will require but little work. The trail, keeping now altogether on the north side of the Teton river, reaches in seven miles the ascent to the low prairie plateau, and here the wagon road could with advantage strike in a southwesterly course towards Sun river. If that course should be adopted, Mud creek, a tributary to the Teton river, with ample wood and water, would afford good camping places. This creek, which in its easterly branch reaches high up towards the divide of the Sun and Teton river waters, would enable trains to shorten the long journey between those two rivers considerably, indicating at the same time the direction which will have to be followed to strike the place where the two rivers come closest to each other. Here two large prairie buttes on the top of the divide would serve as good landmarks, and, leaving the most easterly one of the two to the west, the right course could not now be missed. Sun river, at this place, about twenty-eight miles from the Teton, would be reached at a spot where a creek empties into it about four and a half miles below the Indian agency. It it is said that no coulées intercept the direct course of the above-described road, and that the ground is favorable in every respect. Sun river could be crossed at the point where it is struck, and followed up to the farm and crossed. Trains reaching here would be obliged to take in wood, because the rest of the timber to be found would only be on Dearborn river, which is thirty and two-tenths miles distant from the farm. The whole distance from Sun river to Fort Benton would by the above mentioned route amount to fifty-seven and three-quarters miles, being thirteen miles further than by the prairie lake; but good water, plenty of wood, and convenient camping places would be secured to the road. Unfortunately for us, our guide took us to the westerly branch of Muddy creek, which we struck in two and a half miles from the ascent to the prairie plateau. Made ten and a quarter miles, found sufficient wood and water and good grazing.

December 19.—Started at 10 a. m., and took our course southward, to pass between the two high buttes which crown the dividing prairie ridge. A gradual ascent brought us in three hours to that place, and, finding that the distance to Sun river was considerable, we pushed on with the pack train, Mr. Irvine remaining with the wagons. After crossing three deep coulées, we arrived, 5½ p. m., at Sun River farm, having marched at a brisk gait for the last sixteen miles. A cold and strong wind set in early in the afternoon. Made thirty-two and nine-tenths miles. Mr. Irvine having left the wagons, on account of darkness, at the first coulée, arrived with the animals at 8 p. m.

December 20.—Animals started out for the wagons in a heavy snow storm, but returned after an unsuccessful march.

December 21.—Sent a man over to the Teton mission to obtain a cart, as the peculiar construction of the cañon does not allow of its being carried any further without the risk of injury to our mules. A renewed attempt is made to bring the wagons in; they were found, but plundered by some Indians.

December 22.—The cart came in, and we caught up all the animals to make an early start. Mr. Dawson and my party number eight men, with thirty-nine animals altogether.

December 23.—Some animals that had run off during the night delayed our start until 11 a. m., the day being clear and warm. Sun river is still frozen over. At noon a west wind sprang up, augmenting hourly in strength. Made thirteen and two-tenths miles to-day, and camped on Bird-tail Rock creek. Snow twenty inches deep.

December 24.—Started at 10 a. m; clear and warm, but a strong southeast wind blowing almost a gale; snow is getting deeper and deeper the nearer we come to the little divide, on the northeasterly side of which we found it five or six inches on a level, with drifts of many feet in depth; on the south side eight or nine inches of snow, and still increasing towards Dearborn river, at which stream we encamped, having made 18 miles

December 25.—Started at 9 a. m; cloudy and warm. The snow is increasing in depth at every mile we make, and at Little Prickly Pear we find it from 12 to 15 inches on a level. The drifts are almost impassable for a vehicle, and our animals are getting very tired. I am obliged to cache the cart at the south side of the lowest Prickly Pear crossing, where I camped; made 14¾ miles.

December 26.—Started at 10 a. m; our animals had poor feed last night; cloudy and warm morning; clear towards noon. The crossings in the cañon are partly frozen, making them difficult; snow from

16 to 20 inches deep. No trail is broken and our animals have hard work. On the top of Medicine river divide we found 26 inches of snow, and camped on Little Prickly Pear creek, one and a half miles higher up than I did on my eastward trip; made $15\frac{5}{10}$ miles.

December 27.—Started at 10 a. m; clear and cold morning. The snow is badly drifted and has, on a level, a depth of 26 inches; we are travelling very slow. Found a very comfortable camp on Soft Bed creek; made 11 miles.

December 28.—Started at 10 a. m; cloudy and cold day; we have been threatened all day long with a snow storm, but it cleared off towards evening, turning, though, very cold. The depth of the snow to-day is fast decreasing. At our camping place, on Big Prickly Pear creek, we found only three inches; made $13\frac{5}{10}$ miles.

December 29.—Remained in camp, which is about $2\frac{1}{2}$ miles from the top of the divide; a heavy wind storm with driving snow prevailed until noon; when the snow stopped falling one of the men went out to examine the condition of the road and reports little snow on both sides of the divide, but a deep snow-bed on its top, with a heavy crust, strong enough to bear a man.

December 30.—Started at 10.15 a. m; wind storm with snow; the snow towards the divide is 15 inches deep; on the divide a drift of 80 yards in length, and five or six feet deep; the crust will bear a man, but our horses sunk through, and have hard work. On the west side of the divide I found only six inches of snow, with occasional drifts of the same depth down the cañon of the north fork of the Blackfoot river; near the junction of the north and south fork we struck a well-beaten trail, leading towards the low Hell-Gate pass, which is a very fortunate circumstance and relieves our animals considerably; we camped on the highest Blackfoot crossing, known as Belknap's camp; made $15\frac{5}{10}$ miles. The Blackfoot is frozen across at this place, but higher up the ice broken by the falling of the water.

December 31.—Started at 10.30 a. m; clear and cold day. The snow is from six to eight inches deep; the beaten trail helps us along beautifully. Found both the lower Blackfoot crossings frozen over at camp; the snow is from four to five inches deep; made $14\frac{5}{10}$ miles, and camped one mile above John Grant's cabin.

January 1, 1860.—Remained in camp to give a rest to the animals; warm and cloudy day.

January 2.—Started at 9.15 a. m; cloudy and warm. Found the Hell-Gate river crossing frozen over; three miles lower down I recrossed the river, and kept on its right bank for 10 miles, and joined the party only at Adam's corral; camped on Flint creek, having made $20\frac{1}{4}$ miles. Flint creek had partly overflowed its valley, covering it with a sheet of ice; snow from four to five inches deep.

January 3.—Started at 10 a. m; cloudy and very warm day; the trail in some places entirely covered over with snow, and at other points it is drifted to a considerable depth; made the 12th crossing on the ice, and avoiding the 10th and 11th by passing the rocky point at the water's edge; found the 9th crossing only partly frozen over, and too deep to ford; and it being too late to take to the high water trail, camped at the river bank in 13 inches of snow; we made 19 miles.

Towards evening a wet snow commenced to fall; we feared that it might turn to rain. Heavy snow drifts forced us to make a detour.

January 4, 9.15 a. m.—Clear and cold; the water has fallen during the night, and we made the ninth crossing after having removed the ice; eighth crossing frozen over, but the ice covered by fifteen inches of water; all the rest of the crossings, with the exception of the two lower ones, we found frozen over, but found in following the beaten trail that most of the regular crossing points had been avoided, and even the road for some time located over the high water trail At 11¼ a. m. it began to snow heavily and continued so until evening.

The Big Blackfoot river was crossed on the ice, but higher up and near the foot of the plateau, and encamped on White Thorn creek, having made 23$\frac{5}{10}$ miles. It ceased snowing at 8 p. m., but recommenced at 10 p. m., and fell to the depth of 6 inches.

January 5 —Started at 10.15 a. m ; it had been snowing all night, and we started in a snow storm. I received last night by the Pend d'Oreille chief, Alexander, the first information that you and your party were obliged to build winter quarters on the St. Regis Borgia river. The trail is much blinded, and it still continues to snow heavily. I crossed Hell Gate river for the last time, at the right of the cañon and above Rattlesnake creek, on the ice. It cleared off for a while at 12¼ p. m., but soon recommenced to snow. We found the lowest Bitter Root crossing open for six feet on the easterly side; made 10 miles, and encamped on the Lo-Lo's fork.

January 6.—Started at 9 a. m.; it had been snowing all night, and it blows now strongly from the south with a heavy snow storm. The trail is entirely filled up by wind and new snow. The upper Bitter Root crossing we found open; we reached Fort Owen at 1½ p. m., having made 16 miles. My animals were all in poor condition, and the depth of snow in the valley promised but a scanty chance for food for them, but I trust that after a week's rest I shall be able to start with some of them for your camp. With regard to my trip I am sorry to say that I have not been able to carry out fully your instructions.

The intense cold prevented me from running a compass line, and the depth of the snow obliged me to abandon the wagons in order to save my animals. Your later orders, to examine the road *via* "Lewis and Clark's Pass" and the Big Blackfoot river, I only received on my return to this place. By information from several persons, I have learned that as yet the unexplored roads lead towards the headwaters of the Big Blackfoot river, and both could be made practicable for wagons with but little work. The first one leaves the Hell-Gate river one and a half miles above the mouth of the Big Blackfoot river, strikes over towards the camas prairie, and keeping along its edge gains the Big Blackfoot valley at a point below Lander's Fork. This road, which is certainly the shortest, must avoid all the Hell-Gate crossings, and, as it is stated, offers only one difficulty, consisting in the first ascent after leaving Hell-Gate river. The second road leaves the Hell-Gate valley at Flint creek; the ascent of the bordering ridge would require some work, but it is by far not as steep as the one by the lower road; and then, as my informant assures me, that only a little side-hill work in a few places and an occasional cutting of a few trees would render the road practicable for a wagon road.

I am, sir, very respectfully, your obedient servant,

P. M. ENGLE,
Topographical Engineers.

J.

CANTONMENT JORDAN,
Bitter Root Mountains, W. T., February 15, 1860.

SIR: I have the honor to submit the following report of my journey, undertaken pursuant to your verbal instructions of May last, to examine a route for a military wagon road across the Bitter Root mountains, from the Smakole prairie, on an east line, to the Hell-Gate defile. Furnished with such information of the route as given by certain Cœur d'Alène Indians, and as described by the missionary, the Rev. Father Joset, and with the hope and view of obtaining their assistance, I started from the Dalles without delay.

On the 3d of June I left the Des Chutes Landing, on the steamer Colonel Wright, and landed at old Fort Walla-Walla on the 5th, and reached the new fort on the 6th, where I found Mr. Kollecki in charge of the astronomical observatory, and who informed me that he had already despatched an Indian expressman to the Cœur d'Alène mission for the necessary guides for my trip.

While at Walla-Walla I endeavored to secure the services of some Nez Percés Indians to accompany me to the Hell-Gate, but when I mentioned to them the particular route that I desired to examine, they represented to me that the whole region was an immense bed of rugged mountains, and over which they never heard of any person having travelled, and therefore they declined to join my party.

While at Walla-Walla the Rev. Father Joset arrived, together with Agustine, a Cœur d'Alène Indian, who had been sent for as one of my guides.

This man declared that the contemplated trip could not be undertaken yet, because the streams were too high and the snows in the mountains too deep, and that it would be impossible to pass over the more difficult regions before some time in August. I insisted upon starting immediately for the examination, when Father Joset informed me that the other two guides who were necessary for the full examination of the route had not yet been obtained, and that it was, indeed, doubtful whether they could be obtained. It evidently seems that the route in question, from the accounts of all, had not been a travelled one, and that no one knew the route continuously from the waters of the Snake river to those of the Bitter Root.

It would seem, however, that these three were three Indians who knew the three sections of the route. These were Ko-ne-moo-say, Koahlis, and Agustine. The latter knew the more difficult mountain sections and the higher ridges. He told me that he had discovered this section while hunting, and that he came in view of the Bitter Root valley about twenty-five miles below the Hell-Gate defile, but that he did not descend into the valley.

Father Joset told me that the other two guides were absent for reasons that I might account for.

Agustine was apparently uneasy and displeased, first, because he disliked to show the country of his people, fearing that they would lose it; nor was he at all anxious to undergo the fatigues and difficulties of the trip; but when he found that the absence of the other two guides was sufficient to deter me from travelling, he mustered his best will for the undertaking.

He desired to go and take leave of his family, promising to meet me at the camas plains of the Cœur d'Alènes, on the 20th of June, when he would join me and guide me from there to some of the higher peaks of the Bitter Root mountains, in the country of his tribe, from which, though probably covered with snow, I could have an extensive view over the mountains and the country generally, and saying "then you can see all and satisfy yourself."

As there was a probability of finding the other two guides at this same camas prairie, I decided to let him go, with the understanding that he would meet me on the 20th of June.

While the reconnaissance from the Smakole prairie was thus retarded on account of my Indian guides, I determined to explore the line from the point where the military road crosses Snake river to the Smakolde prairie. The map of this exploration has been compiled in the office this winter. In order that you may know the character of the country in detail, I make the following extracts from my itinerary, which were made in the field at the time:

June 14.—I ferried Snake river, and encamped at the mouth of the Pelouse; grass and wood scarce. My party consisted of Toohill, Frank Hall, a half-breed, and Slougharehy, the Pelouse chief.

June 15 was cloudy and windy. We crossed the Pelouse three miles above its mouth, and ascended a high and steep bluff to the table land; and from the point where your wagon road joins this trail we travelled, magnetic, north 24° east, and encamped on the Pelouse river, at the point where Colonel Wright encamped when holding a council with the Pelouse, in September, 1858. Our march was eighteen miles; grass and wood in abundance.

June 16.—At 8 o'clock a. m. the weather was clear, with a temperature of 66° Fahr. We travelled up the Pelouse, on the left bank, for seven miles, when we struck the Smokel creek, and followed it. It flows through a valley of easy grade, and bounded on either side by high, rolling prairie, and in a generally east and west direction. The valley at its mouth is somewhat narrow, but in a quarter of a mile it widens, and in five miles it runs out into the general level of the country. In eight miles from the river it is closed in again by slopes one hundred and six hundred feet high. The soil in the first ten miles is somewhat gravelly. Further up it is somewhat richer, and in fifteen miles it bears the camas and other roots upon which the Indians subsist. In the first ten miles there is little or no wood, but higher up the cotton-wood, aspen, birch, and brushwood fringe the banks, more or less.

At the mouth of the creek the water is some eight feet wide and six inches deep, and supplied mostly by a small tributary coming in from the south. But the creek is dry in many places, and hence, at times, water might be scarce, though there are many springs that might be improved. Grass is good along the whole route and, indeed, the road is good, requiring but little grading along the valley of the Pelouse for wagons to pass. This day we travelled twenty miles up the creek, twenty-seven miles from last camp, and encamped on the creek.

June 17.—Thermometer 52° at 5 a. m. Continued to follow up the valley of the Smokel creek. The valley is dotted with pines, and in many places it is cut up by narrow sloughs, now dry, but they would interpose no serious obstruction in the construction of a wagon road. In twenty miles our road joined a heavy trail coming in from the southwest, where it leaves the valley of the Smokel creek, and leading over a rolling prairie region in eight miles it reaches a good spring of water, which is about two hundred feet to the north of the trail, on a side hill. In three miles more it reaches the bottom of the Smokel, which we ascended for two miles, and encamped at a spring near a bunch of thorn-berry bushes, distance from last camp $34\frac{2}{10}$ miles. The creek winds through a flat which is a quarter to one mile wide, and which in early summer, being under water, produces the camas.

June 18.—We followed up the Smokel prairie for seven miles, and encamped on the creek of the same name, and at the point where it joins to the northwest foot of the Tathunah hills. These hills are from twelve hundred to fourteen hundred feet high, and some three miles long. The top of the southern portion is sparsely timbered with pine; there is a ravine near at hand which is also timbered, and where a spring of water offers a good camping ground. The country to the south and west of the Tathunah is a rolling prairie, extending as far as the eye can reach. On the north and two miles distant is a mountain ridge covered with pine. The intermediate region is low and diversified by small spurs and prairie bottoms, which last connect with the Smokel prairie, and it is through this gate that the road would pass on approaching the Bitter Root mountains from the west.

Towards the east the country for twelve miles is an undulating prairie, with scattering pines, when the westward foot slopes of the Bitter Root mountains are reached. These rise to the east and northeast in lofty peaks and spurs, and covered with dense pine forests, while in the southeast they rise in masses of broken flats and free from timber. I ascended the northern peak of the Tathunah to obtain a view of the distant ridges, but found them obscured by the haze of the atmosphere.

June 19.—I despatched Slougharehy, the Pelouse chief, on an express to your camp, forwarding by him a map of the line travelled and examined up to date, whence I proceeded to the camas prairie in search of the Cœur d'Alène guides. Leaving the Tatunah hills, we passed over a rolling prairie country in the general direction of north 30° west, magnetic, for five miles, when we crossed a ridge five hundred feet high, and steep; in eight miles from Tatunah we crossed a small creek called Ki-ah-ne-mah; four miles more we ascended a ridge nine hundred feet high, and in 1½ mile descended to the valley of the Pelouse river proper—here called the Mocalishiah. The valley is here three-tenths of a mile wide, and timbered with the pine. The river is thirty feet wide, two feet deep, with sandy bottom; its general course is west; at the distance of seven miles further we encamped on a creek where the water stood in pools, here the grass was good and wood abundant.

The weather was exceedingly warm during the day, severe upon our animals, that were not in the best condition; our march this day was nineteen miles.

June 20.—We started at sunrise, it was clear and pleasant, thermometer being at 50°. The road continued over rolling prairie, when in 4½ miles it crossed the Tugossomen creek; in two miles further another small creek; thence ascending, in 1½ mile a ridge of six hundred feet descends, and in four miles reaches the Medenhauld or camas prairie of the Cœur d'Alènes; distance from last camp ten miles. The prairie is about one mile wide, and bordered by mountain spurs covered with pine forests. On a flat elevation, on the left bank of the Medenhauld creek, was the camp of the Cœur d'Alène Indians. It was at this time they were digging roots; bands of horses were grazing in the bottom and along the creeks.

This, then, was the appointed place and time for meeting the guides who were to conduct me in my exploration across the Bitter Root mountains. The guide Agustine was here, who, with other Cœur d'Alène Indians, came out to meet me, and who pointed me to a spring nigh at hand; I encamped.

During the day the chiefs and principal men, who had been my old friends, visited me in camp and exchanged news. I soon learned that the services of my guides had not been obtained. One, Kah-lis, was absent, and Koo-na-moo-ney, who was present, was in a doubtful mood, and told me, as an excuse, that twe could not gonow, as all the guides were not present. They were evidently excited regarding the construction of the military wagon road, and inquired anxiously as to its probable location. Koo-ne-moo-ney said he thought we should take the Smakold prairie routes, but when asked to go as guide he refused. I saw then, from the mood of the Indians, that their services could not willingly be secured, and their disposition was to retard my movements, obstruct my passage through their country, and to have me consume a large portion of the summer in fruitless efforts at an examination. Without guides, without information, and in a region known only by its difficulties, and with Indians in their present mood, it seemed a useless effort for me to attempt to make a direct exploration of this special route that you sent me to explore and report upon. I disliked, however, after all my labor and expense of time and means, to leave the mountains without gaining some approximately correct information that would form the basis of your operation, or subserve the purposes of future explorers. I determined therefore, as Agustine was the more willing of the three to accompany me into the mountains, that I would avail myself of the opportunity, and to attempt to ascend some of the higher ridges of the Bitter Root range, and thus obtain a glimpse or view of the country, and then arrive at some definite conclusion regarding its topographical character and formation. Therefore, on the 21st of June, I left the camas plain with feelings of some anxiety, though determined to do the best with the means and aid at my disposal.

Before leaving camp, however, the old Cœur d'Alène chief, Skah-hah-le-me, called to see me, and said that his people had held a council the night previous, and, passing around the pipe, had remained up late, arguing the question whether Agustine should be permitted to accompany me, or not, to the upper country. The majority prevailed in my favor, though Koo-ne-moo-ney was against it. Agustine was present and heard the conversation, and appeared indignant at the idea that the Indians should interfere with him. He started willingly, though lamenting the difficulties that we should meet with in ascending the ridges, where we would have to clime for days over steep side hills, where the ground was obstructed by dense forests and fallen timber, and even now covered with snow.

Starting out we travelled up the open prairie valley for three and a half miles, thence through open timber for three miles, where we crossed a low spur and descended to a creek, which we crossed and encamped on its right bank. There was good grass in the open timber.

June 22.—Started at sunrise, weather clear, thermometer 42°, (Fahr.;) our route lay through open timber, and in five and a half miles along a side hill, and following up the right bank of a small creek, a tributary to the Nedenhauld, and in one mile more we crossed it at the junction of two branches; thence over a ridge covered with dense forests, fallen timber, and brushwood, and in two and a half miles from our camp we reached a camas prairie, 2 miles long and one-half of a mile wide, closed in by mountain spurs, densely timbered.

We followed a small creek which joins in this prairie a second creek, that rises from the south, and which together form a stream 20 feet wide. We passed along the main creek, through a narrow valley on a good road for four miles, when we crossed it, then the trail leads over a low ridge one mile, and to the north of a ravine that leads into the valley of the south fork of the St. Joseph's river; we crossed the St. Joseph's and encamped on its right bank; distance travelled, 25 miles.

The river here is eighty feet wide and two feet deep, with a gravelly bottom and swift current. The valley is three-tenths of a mile wide and densely timbered; our animals had sufficient grazing along the edge of the river and at places where the timber was less dense.

June 23 —We moved up the south fork of the St. Joseph's, eastward, crossing three creeks and making six crossings of the main river, and encamped on a camas prairie five-tenths by two-tenths of a mile in area. This prairie is along the river and at the foot of the high ridge for which we were travelling.

June 24.—We rested in camp, for our animals were now weary and fatigued, and we thought it best to halt and prepare for our task in ascending the ridge. All had gone well up to this point, but here our evil genius seems to have attended us in the person of Damass, the brother of my guide Agustine, whom we met hunting in the mountains. After a series of endeavors and intrigues on the part of Damass, he succeeded in changing the mood of his brother, Agustine, who then determined not to accompany me further. He came to me and said as follows:

"Let us have a better understanding. I told you at Walla-Walla that there was no road through this section to the Flathead country; the mountains are too difficult. Those high ridges yonder, (pointing to some high spurs of the Bitter Root range,) are small compared to those over which we shall be compelled to pass, and there are many of them. The road passes over high and steep mountains and down in steep ravines and cañons. If all the Americans would work here a thousand years they could never make a road. I think the Pend d'Oreille route is the best for you. There you do not pass over difficult mountains. It was Koo-ne-moo-say who told you that a wagon road could be had here, and he says this because he owns land along the Pend d'Oreille route, and he fears that if the soldiers make a road along that route he will lose his land. I am sorry to see Father Josef and yourself determined to take this route. Those who have told you that a road could be had here have deceived you and made you lose time in exploring. I say, again, that if all the Americans would spend a thousand years here, they could not make a road."

This conversation with my guide showed me that he was determined to go no further, and that all endeavor would be now useless. My animals were jaded, provisions scanty, and even in this extremity I should have tried to continue the exploration further, had I not perceived that these men were planning mischief against me, and laying for me a trap; for they now desired to detain me until other Indians should join them, for they daily expected others from the main body. You have, then, my reasons for returning, and though I did it reluctantly, it was my safest alternative to preserve my small party from the hands of men whom I now could trust no longer.

I returned to the Smakold prairie, where I found encamped the Nez Percés and some other tribes, who were here digging camas. They met me in a friendly manner, but could give me no information relative to the proposed road; but they all regarded it as a matter of impossibility to discover a wagon road south of the one that you had originally proposed following, viz: the Cœur d'Alène Mission and St. Regis Borgia river.

Among the Indians whom I here met was "Three Feathers," whom I had often met in the mountains before. We gathered in his lodge to exchange news, &c. It was here I met the old chief, Yah-moh-moh. The old man, not knowing my exact object in this region, began speaking very freely regarding the mountains. He said: "when young, he hunted in this region and found plenty of game. Water and grass were abundant and travelling good." I then asked for a guide, to show me the section he spoke of, and for the hire of a few horses. Three Feathers refused. "Wait, you will get plenty of horses." I left the lodge and invited them to come and see me in the evening. They did so, and dined with me, when, bringing up the subject of the road, I was only surprised to find that their opinions now were just the opposite to what they had expressed in the morning. The old man made an energetic speech, and declared his friendship for the whites, &c., but described the mountains as formidable, the forests and underbrush impenetrable, and the streams dangerous, if not impassable, and implored me not to think of exploring the route; that if I did, I would perish, and men would say that the Indians had killed me. He thought the route by the Cœur d'Alène Mission the best. The consequence of all this was, that I could obtain neither horses or guides. Never have I seen Indians more impertinent, or unwilling to do service of any character. I could not even secure the services of one of them to bear you an express. I desired Agustine to return with me, but his brother Damass told him that you would hang him. Thus you will readily see the circumstances amid which I found myself, and the peculiar trying difficulties of my position. I then determined to despatch Frank Hall with an express to you, mounted on my best horse, and abide your further instructions; and as you are well aware, with all the difficulties around us, I was still willing to undertake the exploration of the line embraced from the Tathunah to the Hell-Gate defile, which you thought might be practicable. I satisfied myself of one thing: the extreme aversion that the Indians have against any wagon road passing through their country; and I have no doubt but that it affords a constant theme for conversation and discussion among them.

I remained in camp at the Tathunah hills until the 3d of July, awaiting the return of Frank Hall, when I started for your main camp, where I arrived and reported to you on July 7th.

I am, sir, respectfully, your obedient servant,

G. SOHON,

Guide and Interpreter, Military Road Expedition.

Lieut. JOHN MULLAN, U. S. A.,

Commanding Mil. Road Exp'n from Walla-Walla to Ft. Benton.

Respectfully referred to Capt. A. A. Humphreys, U. S. Engineers.

JOHN MULLAN,

1*st Lieut.* 2*d Artillery, commanding Mil. Road.*

CANTONMENT JORDAN,

Bitter Root Mountains, W. T., February 19, 1860.

K.

CANTONMENT JORDAN, W. T.,
February 20, 1860.

SIR: In obedience to your instructions I have the honor to submit the following report of my trip as expressman from Fort Walla-Walla to Cantonment Jordan.

February 5.—Left Fort Walla-Walla at 5 a. m. and made the Touchet river at 4 p. m., when we camped for the night.

February 6.—Started at 7 a. m., made Snake river at 4 p. m.; crossed; four men and five horses pushed on. Camped on the Pelouse, two miles from its mouth. Here I found a roan horse which gave out with me on my October trip. The Indians not giving proper account of said horse I will take him to the Mission.

February 7.—Started at 5 a. m.; made good time, and camped on a camas prairie, (at the mouth of a ravine.) No snow, except on the north side of the hills.

February 8.—Started at 7 a. m.; made the Pelouse river at 4 p. m.; crossed and camped. (Horses nearly swimming.)

February 9.—Started at 8 a. m.; good travelling, and reached Camas creek and camped. Here we found six "Indian horses," which the Indians will drive in advance in the morning, with a view to break the trail through the snow, as far as St. Joseph's river.

February 10.—Started at 7 a. m.; slow travelling, snow two feet five inches; camped at Long Prairie, 9 miles east of St. Joseph's river.

February 11.—Started at 5 a. m.; reached the St. Joseph's at 2½ p. m.; camped; made arrangements to leave four of our horses and hire four horses in place of them; hired four horses, at $3 each, to take us to the Cœur d'Alène Mission.

February 12.—Started at 8 a. m.; crossed the St. Joseph's river on the ice at mile post 166, one mile above the bridge; struck across the lake over to the Cœur d'Alène river, which we struck one mile below the point where the military wagon road first strikes it; crossed on the ice and travelled upon the north side. Camped at the ferry on the Cœur d'Alène.

February 13.—Started at 8½ a. m.; reached the Cœur d'Alène Mission at 2½ p. m.; made preparations to start next morning. Rev. Father Josette furnished 12 lbs. of flour and four loaves of bread.

February 14.—Started at 11 a. m.; made Mud prairie at 4 p. m. and camped at the crossing. Snow two feet nine inches deep. Number of fallen trees, this day, across the road is 46.

February 15.—Started at daybreak; travelled rapidly on the crust of the snow; snow shoes sank in but little; camped at 223 mile post. Number of fallen trees across the road to-day are 57; snow, 3 feet.

February 16.—Started at 5 a. m.; slow travelling; reached the foot of the divide at 4½ p. m.; camped; snow six feet deep. None of the crossings are frozen. Number of fallen trees across the road this day are 55.

February 17.—Started at 5 a. m. and made the summit of Sohon's

Pass at 6 a. m.; seven feet four inches of snow on the summit of the mountain; pushed on to the log-house and camped. Snow here 8 feet deep; snow shoes sank in 13 inches. Number of trees across the road to-day seven.

February 18.—Started at 5 a. m.; dark; bad travelling on account of the melting snow. Arrived at Cantonment Jordan at 3 p. m.

Total number of trees across the road from the Mission to Cantonment Jordan are 165, of which 158 were on the road along the Cœur d'Alène river. But little snow before reaching the mountains; snow increasing gradually to the highest section of the mountains. The maximum depth on the summit of the mountains, as measured, being seven feet four inches.

I am, sir, respectfully, your obedient servant,

P. E. TONHILL.

Lieut. John Mullan.

Cantonment Jordan, Bitter Root Mountains, W. T.,
February 21, 1860.

Respectfully submitted to Office Ex. and Surveys, War Department.

JOHN MULLAN,
1*st Lieutenant* 2*d Artillery, in charge of Military Road.*

L.

Fort Owen, Bitter Root Valley,
Washington Territory, March 7, 1860.

Sir: According to your instructions to examine the valley of Snake river upwards from Fort Taylor to "Red Wolf's crossing," I left your camp, at the mouth of the Tuckanon creek, on the morning of July 1, 1859, with one man and three horses. The Indian trail leading to the Wolf's, on Al-pah-hah crossing, follows the north bank of Snake river, which, offering less difficulties for horses, I crossed the river at your camp and took the above-mentioned trail, following it upwards to the place of my destination. Snake river valley, as far as I ascended it, is entirely destitute of timber, and the enclosing bluffs are also naked, and only covered with bunch grass, giving it quite a mountain aspect. The valley itself gains gradually more the character of a cañon, and the bluffs become in the same degree more rocky and perpendicular, and affords more difficulty to the location of the trail. At the top of the bluffs the surface gains the form of long rolling plateaus, cut up by steep ravines, which extend backward from the river from four to six miles, many of which, in spring, contain running water heading in the plateaus. These tributaries to Snake river are not very large, but are of great importance to the Indians, who by this means irrigate their fields, which they have established at every little point where the valley extends and the soil promises to yield a crop. The Indians are so anxious to profit by every piece of ground fit for cultivation, that islands, and even portions of islands, are used for this purpose. Besides wheat and corn, they raise vegetables of different kinds, and gain sufficient crops to encourage them in their labors. The soil generally does not appear to be rich in the valley, but gravelly and sandy; but the plateaus on both sides produce fine grass, offering magnificent pasture grounds. The cultivated ground, amounting to from three to four hundred acres, is distributed into seven farms, of greater or less extent, the lower two of which belong to the Pelouse Indians, and the remainder to members of the Nez Percés tribe. The river flowing between "Red Wolf's crossing" and Fort Taylor is a semi-circle, convexed towards the north, bends little in itself, but, where it does, forms sharp curves, occasioned by projecting rocks springing out from the bluffs. The average width of the stream is three hundred and fifty yards at the stage of water in July; and I noticed only one place, about thirteen and a half miles above Fort Taylor, where, for a mile and a half, it reaches a width of five hundred yards. The current of the water is very rapid, but I am unable to give its velocity; I think, though, that it is about the same as observed by Mr. Weisner at Fort Taylor. No obstruction of any consequence exists to its navigation for the whole distance of sixty-five and a half miles, and the only rapid which extends from bank to bank—situated about four and a half miles above Fort Taylor—is far more insignificant than all those below the mouth of the Pelouse river. At four other points, all situated above the just mentioned place, the river forms small rapids, extending partly across the stream, leaving a large space of water unobstructed for shipping purposes. Snake river, in comparison with other large streams on this side of the Rocky mountains, forms but few islands; the only one of any extent on the portion of it seen by me is twenty-five and a half miles above Fort Taylor, and just below the first Pelouse farm. This is an island of one mile in length and half a mile in width, covered with luxuriant grass, used by the Indians for pasture ground when the side hill grass has been dried up by the summer's heat. The main channel of the river passes the southerly shore of the island, which, in fact, is only divided from the main land by an arm of water not exceeding twenty-five yards in width; besides this largest island, I counted thirteen others, all of which, but one, are above the lower Pelouse farm. The depth of Snake river must be great, and at all stages of water above fording. Its bed is gravelly and in some portions rocky. Below Alamoté creek, which is thirty-nine and a half miles above Fort Taylor, five small creeks empty into Snake river from the north and two from the south; six miles above Alamoté creek, where the upper Pelouse farm is established, the second large creek, named "Awanwi," empties into Snake river just above the first Nez Percés farm. From the west bank of the Awanwi creek the Nez Percés race ground extends along Snake river for one mile and three-quarters. This is a level but gravelly tract of land, with a soil too barren to produce even sage brush. From Awanwi to Skalaisson's creek the distance is fifteen miles, in which interval four more small creeks empty from the north side into Snake river. Skalaisson's creek, running in the ravine which the trail from the "Red

Wolf's crossing" towards the north follows, heads four and a half miles from its mouth, which last is about two and a half miles below the mouth of the Al-pah-hah creek. At Al-pah-hah creek a tribe of Nez Percés, under their chief, Timothy, have a permanent village, with houses constructed of buffalo skins and mats, supported by a light frame-work. I was friendly received by Timothy's son, Edward, ferried over, and treated to salmon, which fish we caught in great numbers in the waters of Snake river. Opposite the mouth of Al-pah-hah creek is a high, rugged, and timberless mountain, called "Tai-luts," which overtowers all others. This mountain, standing in the course of Snake river, has forced that stream out of its way, causing it to make two abrupt turns to get around its base. At the mouth of Al-pah-hah creek is a spring, known to the Indians as Al-pah-hah spring, which is said to contain some healing power. Snake river being, as above stated, entirely destitute of timber, the Indians living along its shore have to depend entirely for fuel upon driftwood, which, at every freshet, is deposited in great quantities by the water along the river shores, and then carefully gathered by the Indians. The valley of Snake river is so cañon-like that it would be very difficult to construct a wagon road in it, for that reason alone; but then the absence of timber would almost make it an impossibility for railroad purposes. These difficulties would be increased by the sudden sharp curves, which would involve long rocky excavations, or the construction of bridges across the wide and rapid stream. Wagons proceeding to the upper Nez Percés country will find, though, an easy road along the north plateau, if they will keep far enough from the bluff's edge to head the ravines; water and grass are plentiful, and also wood in patches. Leaving "Red Wolf's crossing" without being able to procure a guide, on the morning of July 4, I resolved to follow the plateau road towards the Pelouse river, even without assistance, rather than return with my tender-footed animals over the rocky river trail By information, I had learned of a trail leading in that direction, and concluded to hunt for it and return by it to your camp. The trail in question gained the plateau by the Skalaisson's ravine, which is four and a half miles long. The trail in the ravine is bad in places, but, compared with the generality of mountain trails, is a good one. In four and three-quarter miles I gained the top of a small prairie ridge, and in one mile and three-tenths further I struck a broad trail forking towards the west, which I thought was the one spoken of by the Indians After following it for nine and one-fifth miles, and passing four fine springs and a trail leading from the river towards the north, the trail I was travelling gave out, when I concluded to turn and follow the trail which I crossed three and two-fifths miles back, in search of Mr. Sohon, whom I knew to be encamped at Tathunah hill. It was $2\frac{1}{2}$ p. m. when I lost the trail, and I had to travel fast in order to reach Mr. Sohon's camp before night, as my provisions were almost exhausted, having started with only two small loaves of bread and a few pounds of boiled meat, which last was spoiled by the heat on the first day out from your camp. In ten and a half miles I crossed a prairie ridge four hundred feet high; and six and a quarter miles further I struck a small grove, which I recognized as a camping place of Mr. Sohon, on Sma-kol creek, as indicated on his sketch, of which I had a tracing with me. Still pushing on, I crossed, in three miles and seven-tenths, Sma-kol creek, after having passed acamas prairie. From the crossing I went over a small prairie ridge and camped, after a ride of forty-one miles, on a creek which contained bad-tasting water in holes. No bush or sage brush was to be seen to which our animals could be tied, and we had to turn them loose. Mosquitoes and gnats were innumerable, and kept our horses moving all night long, so that they had disappeared the next morning far out of sight, and were only found after a search of many hours. Knowing now where I was, I determined to take Mr. Sohon's road along Sma-kol creek, and try to reach your camp that night, without hunting up the camp of Mr. Sohon, who might have moved from Tathunah hill. Following down Sma-kol creek on its right bank for five and five-tenhs miles to the crossing point, I struck, in half a mile more, the trail travelled by Mr. Sohon, and reached, in eleven miles and three-fifths, Smokle Creek valley, which I followed down for ten and five-tenths miles. At this spot I found an Indian lodge, and obtained, for a few rounds of ammunition, some tea and salmon. Resuming our march on the afternoon of July 7, we camped on the Smokle creek, after having descended it for fourteen miles and seven-tenths; the day was very hot, and my horses quite fatigued, having made forty-two miles. The next morning I started at daylight, and joined your camp on the Pelouse river at 10 a. m., after a march of nineteen miles. The valley of Smokle creek having been described in Mr. Sohon's report, I think it unnecessary to refer to it here.

Very respectfully, your obedient servant,

P. M. ENGLE,
Topographical Engineers.

Lieut. JOHN MULLAN, U. S. A.,
Com'g Fort Walla-Walla and Fort Benton Mil. Road Expedition.

M.

Report of a reconnaissance of the country between In-chaty-kahn camp and the lower portion of Spokane prairie, and of the portion of Spokane river between the Cœur d'Alène trail and the Colville wagon road ferry.

FORT OWEN, *Bitter Root Valley, W. T., March* 8, 1860.

SIR: Receiving your instructions of September 19, 1859, respecting an examination of the country between some point westward of Power lake and the Spokane prairie, with the view to the location of a wagon road, I was delayed until the 24th of September by continued rain storms at the main camp of the party on the Ten-mile prairie on Cœur d'Alène river, at which date I started with one man, two riding and two pack animals, and camped on Cœur d'Alène river six and five-tenths miles west of the Mission. I had to remain there the next day, my animals having strayed off, being found only late in the evening. On September 26th I reached St. Joseph's river crossing; at evening it began to rain, and continued to do so for the night and next day, obliging me to remain in camp. On September 28th I crossed the St. Joseph's river, and followed the wagon road towards In-chaty-kahn, at which place I camped. The Cœur d'Alène Indians, who were encamped along the St. Joseph's river, annoyed me considerably, in stopping me repeatedly with the view of obtaining some presents. September 29th commenced with rain. I started early, and followed the In-chaty-kahn Creek valley downwards for six and three-fourths miles on its right bank; there crossed the creek, and kept the left bank for one and a half mile, at which point I made two crossings but a few yards apart from each other. The direction of the valley to this point is S. 86° W. No obstruction of any description would offer itself to the location of a wagon road. The low prairie hills which, for the first one and a half mile, closely hug the creek, recede after that distance, and leave a valley bottom of one-half mile in width. The creek, lined with brushwood, contained water only in holes; but the winding bed is cut deeply, showing in different places the sign of a strong water current in high-water season. Large, open pine timber covered the low hills, which produce fine bunch grass. The above-mentioned first crossing of the creek cannot be avoided, as the creek makes a bend towards the north and keeps close to the hills, from which its waters have washed off the foot-slopes, transforming them into a perpendicular cut of ten feet in height. The two following crossings can be easily turned, occasioning only a detour of eighty to one hundred yards. In one and a quarter mile lower down, the road would enter the open timber, and one-half mile from there a crossing to the right bank would be necessary at the same point where the trail crosses, and then follow that bank for one and three quarters of a mile to a place where the trail leaves the In-chaty-kahn valley, and takes a more northwesterly direction. At the place where the road leaves the valley, a small ridge of one and a half miles in width, with easy ascent and descent, lying between two parallel running tributaries of the In-chaty-kahn creek, has to be crossed. Here a little timber cutting will be found necessary, principally on the descent. In-chaty-kahn creek, a tributary of Lahtoo creek, is nine or ten feet wide, and has at the point selected for crossing a gravelly bed, which at the present season is scarcely covered with water. From the foot of the above-mentioned little ridge the road would lead for three and a half miles over undulating ground, covered with open pine timber, to a third tributary of the In chaty-kahn; and in two and a half miles further, over a country of the same description, would strike a fourth one, which runs in quite a broad bottom; part of which is miry, and would require some corduroy work. Three-eighths of a mile from this point the trail forks; the wagon road would follow the north fork to save an angle made by the left hand trail, which strikes the five tributaries of In-chaty-kahn creek one and three-tenths of a mile below the crossing point of the upper trail. From the forking point the trail strikes the said tributary in two miles, and the lower one in one and nine-tenths of a mile. I followed the lower trail to reach camp as soon as possible, as it had been raining all day, and we were drenched and cold. This little water-run affords good camping places, with plenty of water, wood, and grass. The valley is from one hundred and eighty to two hundred yards wide, enclosed by low hills covered with pine timber; the creek itself being lined with willow and cotton-wood. Made 22 miles to-day.

September 30.—Ascending the valley for one and one-tenth of a mile, I struck the point of the upper trail, and one-quarter of a mile from there left the valley, ascending an easy divide of 250 to 300 feet, the top of which I gained in one mile. The descent is as gradual as the ascent, but somewhat more densely timbered; but by removing a few trees any wagon could be brought down to the creek, which flows one and one-quarter of a mile from the top of the divide in a narrow valley. At this point the road has to descend the creek in keeping on its left bank to a place where the pine timber disappeared, there it will ascend the low and gently sloping side hill, and strike, following almost an air line, in five and four-tenths miles, the Spokane river, opposite the termination of the bluff running behind Antoine Plant's farm. From this point the road can follow the Spokane river on its left bank until within five miles of the place, where it leaves the river altogether, and striking over the mountains pass towards the Fort Walla-Walla and Colville wagon road.

No obstruction of any moment will be found to bring wagons over, even in the present natural state of the ground, whereas, on the right bank, same considerable work will be unavoidable. Mr. Kollecki having already reported to you regarding this latter section of the country over which the road would have to go, I think it unnecessary to do so again. The Spokane river, which I crossed at Antoine Plant's house in a boat, I found to be eight feet deep in the channel, and big boulders in its bed. My animal having strayed off, I remained the next day at the farm, and started only October 2 for Colville depot. I camped that night at the Spokane fishery, near the mouth of the little Spokane river, a distance of 20 miles—having taken the most certain route to gain that point.

October 3.—I started early, but, having mistaken the trail, I travelled for about 10 miles in the direction of Pend d'Oreille lake, and only then finding that I was wrong, I had to retrace my steps; but, making a short cut, I gained before evening the Spokane river, about 10 miles below my last night's camp, and, in three marches, arrived at the Colville depot, where I was kindly received by the officers of the post. On October 12, I made a survey of the wagon road from the Colville depot to the Hudson's Bay Company's Fort. This road leads through "Pinkneyville," and then along Morigeau's creek passes the saw-mill, two and a half miles distant from the depot, and following the creek, which winds around a high mountain, gains, in six miles and seven-tenths from the depot, the junction of the lower or low water road, which runs for some distance through Mill Creek valley, and becomes practicable for wagons only late in the season. From the point of junction, the distance to the Hudson's Bay Fort is seven miles and five-tenths, making a total distance of fourteen miles and two-tenths, from place to place, by the upper road. This wagon road is a natural one, and I doubt whether any work has been done on it, with the exception of corduroy bridges, which were constructed during the summer of 1859. At Colville I was hospitably received by the gentleman in charge, Mr. McDonald. In returning, the next day, I extended my sketch a little by going to the old Colville Mission, and to the Kettle Falls of the Columbia river, and then following the wagon road to the grist-mill, which is located on Mill creek, about a mile and a half above its mouth, at the forking of the lower and upper roads. I took the bottom road, and, having run a compass line over the whole distance, I found the lower road two miles and one-tenth shorter than the upper one. At the Colville depot I learned that Commissioner Campbell, of the northwest boundary survey, with party was shortly expected, and that there was a possibility that he might go down the Columbia river. I therefore determined to await his arrival, knowing that you were anxious and intended to have that important stream surveyed from Fort Colville to Fort Walla-Walla. Mr. Campbell, who arrived on the 16th, found that the preparation of the boat would occasion a long delay, and preferred, for that reason, to make the trip by land. Having obtained permission from Mr. Campbell to travel with his company, I started with him and Mr. Warren on the 20th of October. In three marches we reached the lower end of Walker's prairie, where Mr. Campbell was unfortunately delayed for a day, by the straying off of his team mules. I pushed on and camped on the Spokane river, three miles above the ferry. I took advantage of my presence in this part of the country to run a compass line from the ferry to the place where the lower Cœur d'Alène trail to Colville leaves the Spokane river—knowing that this section had never been surveyed. I found the distance between the two points seven miles and three-tenths; of which — miles are rendered difficult by the closing in of the side hills; but just above the ferry a company has commenced to build a bridge, which was not completed at the time I passed. The day was so foggy and rainy that I hardly could see across the river, which runs in quite a narrow valley walled in by steep, rocky hills; open pine timber covers the valley to the very edge of the water, which has a swift current over a rocky bed. Having made 26 miles, I camped on a small creek, three miles west of Antoine Plant's house; at this latter place, which I reached on the morning of October 25, I took an Indian to guide me to the Cœur d'Alène Mission, where I arrived on the evening of October 28. Reverend Father Garroli kindly offered me the accommodation of the Mission, and supplied me the next morning with oats, which I needed for my weak animals—knowing that I would find no grass between the Mission and your camp, which I was informed by the reverend father had been pushed to the east side of the "divide," making two more camps on this portion of the road. I joined your party, October 30, on the "Five-mile prairie," on the St. Regis Borgia river.

Very respectfully, your obedient servant,

P. M. ENGLE,
Topographical Engineers.

Lieutenant JOHN MULLAN,
U. S. Army, Commanding Military Road Expedition.

N.

Report of a reconnaissance from the Cœur d'Alène river to Thompson's prairie, on Clark's Fork; thence to the St. Ignatius Mission; thence along the left bank of Clark's Fork to Horse Plain; thence across the mountain to the Bitter Root river; thence up the St. Regis Borgia river, and over the high divide to the ten mile prairie on the Cœur d'Alène river.

FORT OWEN, BITTER ROOT VALLEY,
Washington Territory, March 9, 1860.

SIR: Having received, on the 18th of August, 1859, your instructions regarding an exploration of a road across the Bitter Root mountains towards Thompson's prairie on the Clark's Fork of the Columbia, which had been described by different persons as probably practicable for wagons, I left your camp at the Cœur d'Alène Mission on the same evening, accompanied by two Cœur d'Alène Indians, with three riding and two pack animals. At the four mile prairie of the Cœur d'Alène river, where your working parties, under Messrs. Williamson and Spengler, were encamped, I joined the party of Mr. Sohon, who started for an exploration of the country towards the Bitter Root valley, and kept company with his party until the morning of August 21, when I parted with him, leaving the Cœur d'Alène river trail about twenty-five miles east of the Mission, and, following my Indian guide, we took a northerly direction, ascending a tolerably steep ravine, with running water, and enclosed by high, thickly-timbered ridges. The little creek, which is crossed and recrossed many times by the scarcely visible trail, heads three miles from the place where I left the Cœur d'Alène river trail, and one mile to the northwest from the point where my present trail leaves it, which ascends a very steep ridge densely covered with pine trees. In a mile and three-quarters I gained the summit of the ridge, which has an elevation of 1,500 feet above the Cœur d'Alène river. To the right of the trail runs a narrow ravine connected by a low saddle with another running down the north slope of the ridge; those two ravines could be used for the location of the road; but the removal of rocks, and of the fallen and standing timber to the mouth of the narrow ravine, would occasion considerable labor. At the top of the ridge I saw two valleys before me divided by a high ridge, both of which contain running water, which joins at the termination of the ridge. In four miles and a half, over a very dangerous trail along the steep mountain slopes, I gained the first creek, which the trail follows upwards; after crossing the creek, which is ten feet wide, I took the opposite mountain; commencing here the ascent, all signs of a trail disappearing, we had frequently to use the hatchet to open a way for the animals, and finally had to cut the entire road, the young timber being so dense that not even a man could have passed through. In this laborious manner I pushed on until late in the evening, having made but two miles and a half since I crossed the last mentioned creek, but gained an altitude of about 2,500 feet. I camped on the mountain slope in a large patch of huckleberries, which proved to be quite an acceptable substitute for water; I was forced to tie up my animals to have them at hand early in the morning.

August 22.—Started at 5.10 a. m. Ascended 700 feet higher, using the hatchet again for a half mile, then followed along the slope of the ridge about 500 feet below the backbone of it. Rock slides, fallen timber, and slippery grass made this an exceedingly difficult road. In a mile and a half we found a little spring on the hill-side, and my Indians proposed to camp, to which I gladly acceded. It appeared to me that the Indian whom you had employed as guide was a little uncertain about the way to proceed. We had only travelled over the trail once in winter time on snow-shoes, and the different appearance of the ground, with its difficulties exposed, evidently puzzled him. Not understanding the language, I could only converse by signs with him, but I easily comprehended that he wanted me to stop to give him time to examine the character of the ground ahead. I furthermore understood that we had to follow the crest of a high, rugged ridge, the southerly slope of which is one rock slide from top to bottom, which lies before me a distance of about six miles. How to bring my animals over was not quite clear. Alexis, the guide, left camp as soon as I had cooked breakfast, and Nicholas, the second Indian, collected the animals, which, at the risk of their lives, were clambering along the rocks to collect their scanty food. The country is filled with smoke and no extensive distant view could be gained, but I could discover in the chasm below me the only place where the wagon road could be located; the fallen timber and rocks are piled above each other to a considerable height, and the creek boils over and through them. The opposite ridge is well timbered, steep, and about 3,800 to 4,000 feet high, and closes in so narrowly with the ridge forming my road that no valley bottom at all is left. Towards evening clouds gathered over us, and it began to rain at nightfall. Alexis returned towards supper-time, and gave me to understand that the road ahead is bad.

August 23.—I remained in camp, as it is still raining. I had to alter the cruppers of my pack-saddles, which injured my mules, and had to make some alterations in the rigging generally. In the afternoon a heavy thunder-storm passed over a portion of the mountain bed, but burst against the highest ridges, and the heavy clouds disappeared almost in the same direction where they come. Alexis starts out again to make a further examination. It cleared off at noon, but the atmosphere is still filled with smoke and haze.

August 24.—Started at 6.47 a. m. The day is warm and clear, but the atmosphere is still filled with smoke. In two miles and three-quarters I gained the top of the ridge, ascending gradually, finding the same obstructions in the road as heretofore; keeping along the very backbone of the mountain, which at places is not wider than two or three feet, I reached in five miles and three-quarters more the highest point, which is a granite rock formation of about 3,500 feet high, with a perpendicular rock bluff to the north, and a rock slide of 60° grade to the south. On the north slope hard packed snow, from two to four feet in depth, fills up the ravines and covers the ground in many places where it has been drifted up during the winter. Four ponds, the largest one at the foot of the highest rock, gave a pleasant relief to the monotonous scenery; a dense pine forest covers all the northerly ridges which extend towards Clark's Fork, forming one immense mountain bed, only the outlines of which can be seen indistinctly through the smoke. To the south the main ridge, which gets steeper and steeper, approaches more and more, though not so rugged as the one I am on. It is much more heavily timbered, however. The next three miles we passed over the worst road I ever travelled in my life; the ridge runs out to a rocky crest, which gives not more than from one to two feet footing; sharp rocks, intermixed with loose stones packed upon each other, made every step insecure, and we had to proceed with great care. For two-thirds of the distance the north slope forms a perpendicular rock wall of 1,100 to 1,500 feet elevation. With hard labor we built a kind of a road for the animals by removing the biggest rocks and by steadying the loose stones in the most narrow places. The descent to the low divide is steep, but no rock slides are in the way, and we gained it in three-quarters of a mile, and encamped on the head spring of the creek which waters the valley. We made only 12 miles to-day, but spent 10 hours on the road; the animals are much exhausted, and so am I; we found good grass at the divide, which itself forms a low saddle connecting the two high ridges.

August 25.—Started at 6.39 a. m. We had to gain the top of the ridge again, as it was found impossible to follow the valley which is rendered impassable by fallen timber and rocks. In one and three-quarters of a mile we reached the summit again, advancing and ascending gradually, and discovered from there that the low hill forming the divide has a width of only one-half of a mile, and that the two head springs, sending their waters in opposite directions, are situated at the foot of said hill. The road on the ridge is much the same as yesterday, and at one point it was so bad that the Indian guide proposed to try the valley once more; we therefore descended, but before reaching the bottom we saw that it was impossible to get along in it, and had to retrace our steps; two of my horses rolled down part of the mountain at the point where we tried to descend, and one of them got hurt badly. I observed two large ponds on the north side of the ridge, and also again an abundance of more. The timber on the ridge we followed is increasing, and the fallen timber augments my difficulties in getting along; three and three-quarters miles from the point where I gained the top of the ridge I commenced the deep descent formed by a long spur running towards the creek bed, which I reached in four and a half miles. In some places the slope was so steep that the animals had to slide, bearing down by their weight the small pine trees and bushes which thickly covered the ground. The creek at the foot of the ridge is 25 feet wide, carries a large quantity of water, and runs in a rocky bed, which is closely hugged by the mountains on both sides; at the point where I struck and crossed the creek a densely-timbered flat borders it on the east side, and over this my way led for four and a half miles. Having made 14 miles I went to camp, as it was getting late, and my animals appeared to be much fatigued.

August 26.—Made a late start, my animals having strayed off during the night. The road to-day leads along the creek, sometimes taking the side hills, but for the main part keeping in the bottom, which is much encumbered by brushwood and fallen timber. I have forgotten to mention that in striking the creek I also struck a small trail, which was so blind that even my Indians lost it frequently during the day. Shortly after leaving camp I crossed a small tributary to the main creek coming from the south, and one mile lower down a second one from the north; three and three-quarters miles still lower a large spring branch, heading in the creek valley itself, forms a bad slough of one-quarter of a mile in width; the foot slopes of the enclosing high ridges leave room to the valley, and the creek winding from side to side forces the trail often over the little spurs. I could not form a good estimate of the altitude of the main ridge on account of the dense smoke which allowed only occasional glimpses of their tops, but I should judge that they were at least 2,500 feet high. I had hoped to reach Clark's Fork to-day, but a heavy thunder-storm which overtook us forced me to go to camp at 3½ p. m., having made 16 miles; an extensive fire was raging on the Clark's Fork, and has extended high up on the creek that I was following. All the grass is burnt, and my animals are badly off for food. At night the mountains presented a magnificent picture—the rain and wind had driven off the smoke, and fire lines, extending high up on the mountain slopes, reaching, in some places, to the very tops of the ridges, and in others spreading with surprising rapidity over the fire-furnishing surface, gave a most fascinating aspect to the landscape.

August 27.—Starting at 6½ a. m., I passed in half a mile a large tributary coming from the northwest. The main creek itself, now a considerable stream, with an exceedingly rapid current, sends its boiling water over huge rocks, which gives the bed the appearance of an unbroken rapid. In one and three-quarters mile below camp I crossed the creek with great difficulty. The water's depth was three and a half feet, and the bottom consisted of one smooth bed covered with a slippery moss which the transparency of the water permitted to be seen most plainly. The high ridges which fall back indicate the approach to a large valley, and also the suddenly changed course of the creek which so far has followed a northeasterly direction, and now suddenly changed towards the northwest. After crossing the creek I travelled for one and a half mile over a high, rolling ground well timbered with pine, which has much suffered, though, from the late fires. I crossed then a small creek which empties two miles below my crossing point into the large creek which I just left. In one and a half mile further I struck the shores of Clark's Fork, after having wound my way through a bed of rocky cones which have a height of from 40 to 50 feet; one-quarter of a mile above the place where I struck the Clark's Fork its current is obstructed by a large rapid, which must be called a waterfall of eight feet. The opposite shore was pointed out to me at Thompson's Prairie. In one and one-quarter of a mile more I found a suitable camping place with some grass along the water's edge. The opposite shore is in full blaze, and the wind blowing from the west drives the flames, unfortunately, in the direction which I have to follow. Several deer are forced to swim the river in order to escape the fire, which extends to the water's edge.

August 28.—Remained in camp. We collected material to build a raft, and before noon we had constructed a kind of floating structure, very fragile indeed, but my Indians think it is more than strong enough to do its duty. The fire is still raging on the opposite shore, but has extended more towards the north.

August 29.—In two trips I landed everything safely on the right bank of Clark's Fork; and after having swam the animals, I took up a line of march along the shore of the river, which has a width of two hundred yards. In three miles I reached a small but rapid running creek, which I crossed at its mouth. On the left bank of the river is a high, almost perpendicular rock connected with the high river ridge, which gradually has neared the river shores by a well-timbered slope of 1,500 feet elevation. The right shore forms a large plain, which, from the water's edge to the bluff, must have an extent of three or four miles; but I could not make an estimation of its width, as everything is enveloped in smoke. In one and three-fourths miles from the creek crossing the rocky bluffs of the right bank close into the river and render the trail difficult for three-fourths of a mile. The road is then easy for four miles, and is only now and then obstructed by thick brushwood; at the end of that distance I crossed a small creek, and two miles higher up I struck a point, which will be well remembered by all who have ever passed it, called the "Bad Rock." It is formed by a rock point running out to the very water's edge, terminating there in a perpendicular bluff. The Indian trail, which, in its present location, offers undeniably great difficulties to the animals, winds high up on the bluff side, and regains in the same serpentine manner the bottom, about three hundred yards above the place where it left it. I am confident that, with comparatively little work, a good wagon road could be made; but it would certainly require blasting for a distance of one hundred and fifty to two hundred yards. In three and three-fourths miles, which were made over an almost level road, with the exception of the last three-fourths of a mile, where the trail passed a few insignificant, low, rocky spurs, I gained a little creek, where I camped, having made sixteen miles.

August 30.—Started at 6.50 a.m. In one and a half mile from camp the low foot-hills of the high river bluffs close into the river at a point where I noticed another river rapid, and which renders the road somewhat rocky for one mile, but offering no serious obstacle. The opposite mountain ridge has been getting lower and lower, and has finally dwindled down to a low hill range. The high bluffs of the right bank recede and give place to a large, extensive flat, which only in some places is covered with open pine forest. Having reached the Kootenay trail, I left the river, and, shaping my course more northeastwardly, I gained the Indian road leading to the camas prairie. This portion of the country having been often described in former reports, I confine myself to the recapitulation of the main ground features. Little or no difficulty will be found to bring the wagons over this portion of the route until the ravine is reached which leads downwards to the camas prairie; there heavy grading for one mile will be necessary, as the hill-sides are steep and the soil is intermixed with rock. Having ascended the steep river bluff one hundred and fifty feet high, I reached in one and three-fourths mile the top of low-timbered plateaus which run along a valley bordered to the northwest by a ridge of 1,100 to 1,500 feet of altitude. The plateau is covered with weeds, sending its waters through the above-mentioned ravine towards the camas prairie, where it sinks into the ground. The plateau itself runs along the foot slopes of the ridge, which gradually increases in height, and finally gains an altitude of 1,200 feet, terminating at the lake, from which point the trail leads over foot-hills for two and three-fourths miles, and then takes, after crossing the creek, to the north slope of the ravine. Of the different trails which lead over the camas prairie, I chose the one which runs along its southwesterly border, which is marked by a timbered ridge of about 1,200 feet. In half a mile from the mouth of the ravine I crossed the creek, and then travelled for six and a quarter miles, over almost level ground, to the top of a small prairie elevation. In one and three-fourths mile further I crossed a creek running in a valley, with easy ascent and descent, and two miles further I gained the valley of Clark's Fork by a gradual and easy descent. At this place I found a well-beaten trail which led to the river crossing; we tried the ford, but found it too deep, and I concluded to ascend the Flathead river to its intersection with the Jocko river, and effect a crossing there. Having made twenty-four and eight-tenths miles, I went to camp a quarter of a mile above the place where I struck the river. My Indians seem to be uncertain in which direction to proceed; but having a map along, I soon convinced them by signs that we had to ascend the valley in view to gain the St. Ignatius Mission.

August 31.—Started at 6.53 a. m. One-fourth mile above camp I crossed a creek which runs in a narrow ravine and heads about three miles to the north on the camas prairie. One mile and a quarter above that creek the river widens to double its extent, and is split up by a great many islands of different size, and well covered with brushwood. Here my Indians tried again to cross, assuring me that the road was much better on the opposite side; but the depth of the water forced us to turn back, and taking up our road again, we had to mount a high river bluff with a steep and rocky descent. Then skirting along the water's edge I gained, at five miles and three-quarters from my last night's camping place, a fine bottom of three miles in length, again hemmed in by the river bluffs. For a mile and a half the trail gains a plateau, which, after rounding a projecting rock, strikes the river at the usual fording place, two miles above the mouth of the Jocko river, in three miles. Here I crossed with much difficulty, and not without drenching my packs. Instead of ascending the valley of the Jocko river, as my Indians intended to do, I ascended the Clark's Fork for one and a quarter mile further, and then followed "Plum creek," knowing that it was the shortest route to reach the Mission. At the entrance of "Mission creek" into "Plum creek" I struck a large lodge trail, which brought me to the place of my destination. I was most kindly received by Reverend Fathers Minetrey and Louis, and hospitably invited to stay as long as I desired. I made to-day twenty-one miles and a half. My animals are very tired, and two of the horses I shall be obliged to trade off.

September 3.—Reverend Father Congiato arrived from Fort Benton, and left.

September 5.—I left the Mission in company with Reverend Fathers Congiato and Minetrey, who intended to visit the Cœur d'Alène Mission. In four miles and a half I struck the Jocko river, and after crossing it, followed down Clark's Fork on its left bank for 16.4 miles, and camped, having made 20.9 miles. The road along the river, with the exception of one point where a mountain closes in for one-fourth of a mile, is a very good one. The ground is level, firm, and not rocky. The small creeks, with gravelly bottom and fine grass along their shores, afford good camping grounds.

September 6.—Started at 6.44 on the trail. After passing the crossing point opposite the place where the road leading towards the camas prairie ascends the bluff, it becomes more and more indistinct, as it is very seldom travelled by Indians. Many rocky and swampy places make the road a difficult one; and a bad rock spur, perpendicular foot slope, which is washed by the river waters, forces the trail over a high and stone-covered mountain, which is almost as bad as the "Bad Rock" above Thompson's prairie. In two places where no ascent to the ridge can be gained, and the river runs close to the bluffs, the trail leads along the river bed with a depth of water from two and a half to three feet. Thirteen miles below the camas prairie trail crossing I struck the Bitter Root river, which appears to flow through a narrow valley enclosed by high mountains. I effected a crossing at the junction of the two rivers without any difficulty, and followed down Clark's Fork seven miles and a quarter further; which portion of the road is good, and only made disagreeable to travel by the dense brushwood. I camped on the shores of the river and on the upper end of the "Horse Plain," having made 20.9 miles.

September 7.—Started at 7.27 a. m., and took a straight, magnetic, southerly course, to gain the top of the mountain ridge which forms the divide of the waters of Clark's Fork, the Bitter Root, and the St. Regis Borgia rivers. The ascent and descent are in some places very steep, and rendered difficult by fallen timber; but the trail is well marked and broad. The distance from river to river I estimated at ten miles and a half, and the height of the ridge at 1,500 feet. Ascending the Bitter Root river valley for eight miles, in which I met with no difficulty on the road, I went to camp, having made eighteen and a half miles. I observed two small rapids in the Bitter Root river, and counted two creeks which both empty on the left bank.

September 8.—One mile above camp I left the river at the crossing point, and striking over the little divide towards the St. Regis Borgia, I camped on this latter stream, after a march of twenty-six miles.

September 9.—I ascended the river valley to the foot of the big divide and camped there, having made fifteen miles. It has been raining all day, and we expect to find snow on the top of the divide.

September 10.—I crossed to-day the divide, and, as predicted by the Indians, found the top covered with snow. From base to base of the mountain I estimated the distance twenty-four miles. Descending the Cœur d'Alène river for six miles, I camped on a small prairie, and had the honor to report myself at your camp on September 11. Knowing that a more careful survey would be made at a later period, when the wagon road would have been located and constructed over the country, I deferred running a compass line from the place where the wagon road will cross the Bitter Root river westward to your camp.

Most respectfully, your obedient servant,

P. M. ENGLE, *Topographical Engineer.*

Lieutenant JOHN MULLAN, *U. S. A.*,
In charge Fort Walla-Walla and Fort Benton Military Road.

O.

FORT OWEN,
Bitter Root Valley, W. T., March 16, 1860.

SIR: Having received on the 29th of July your instructions to examine, in company with Mr. Kollecki, the two forks of the St. Joseph's river, from the point where the military road crosses the same, we proceeded on that day to carry out your orders, and on that date reached the forks of the main river, where we found a camp of Cœur d'Alène Indians. Finding it impossible to travel along the forks on horseback, we made an arrangement with the Indians to take us in bark canoes to the foot of the high mountain, which we saw at a distance of about sixteen miles, but the price which they demanded for their services appeared to us too exorbitant, and we returned to your camp at the crossing of the St. Joseph's river for further instructions. Receiving your sanction to our plan to ascend the river in canoes, Mr. Sohon (whom you had detailed instead of Mr. Kollecki) and myself started July 30 for the Indian camp. Finding three bark canoes in readiness for us, we started, and, after ascending the north fork of the St. Joseph's river for eight and one-fourth miles, we reached a little creek called "Junemassequilam," running around the foot slopes of the mountains, which we intended to ascend the next morning. As this was the nearest point to the peak to be gained by water, we landed, encamped, and prepared for the morrow's journey. The character of the river is the same as observed by you at the military wagon road crossing, the water being deep, with scarcely any current. The banks are low, but steep. The river winds its way through a narrow valley, touching alternately each side of the high mountain spurs, forming in places small prairies, some of which contain lakes, or are cut up by little creeks bordered by berry bushes. The Cœur d'Alène Indians consider this portion of the country the richest berry region in the mountains, and visit it regularly towards the end of July and the commencement of August.

July 31.—Mr. Sohon, myself, and one Indian ascended to-day the mountain, but before reaching the summit we were convinced that we would have no distant view, and therefore retraced our steps. At one p. m. the smoke of the burning timber enveloped the whole country, and our Indian guide assured us that we would have no view until after a heavy rain. In returning the Indian set fire to the woods himself, and informed us that he did it with the view to destroy a certain kind of long moss, which is a parasite to the pine trees in this region, and which the deer feed on in winter season. By burning this moss the deer are obliged to descend into the valleys for food, and thus they have a chance to kill them. We remained during the night at Junemassequilam creek, and returned August 1 in canoes to the Indian camp, and from there on horseback to your main camp on the St. Joseph's river.

Very respectfully, your obedient servant,

P. M. ENGLE,
Topographical Engineers.

Lieutenant JOHN MULLAN, *U. S. A.,*
In charge Military Road from Fort Walla-Walla to Fort Benton.

P.

Report of expressman carrying mail across Bitter Root mountains for military road expedition in April and May, 1860.

April 7, 1860.—Started at an early hour from Cantonment Jordan, and travelled as far as the fourth crossing of the St. Regis Borgia river, about four miles from camp, where we were obliged to put on our snow-shoes for the first time. Snow four feet two inches ; bad travelling. River low at present, but had evidently been high, as it displaced nearly all the foot bridges. Waded eleven crossings to-day; camped at the twenty-third crossing. Snowing all day ; evening clear.

April 8.—Started at 6 a. m. ; travelled very slow ; snow light and dry ; snow-shoes sank in eleven inches. Reached the summit of the mountain at 1 p. m. Snow on the divide seven feet. From this point to the Cœur d'Alène river the snow kept gradually growing thinner. Reached the Cœur d'Alène river at 3 p. m. and camped ; crossings much better to-day.

April 9.—Started at sunrise. Travelling much better to-day ; crossings much worse ; waded seven crossings to the south fork. Here we took to the side hills, but could not travel to advantage, as at some points the bluffs were nearly perpendicular, and had to take the wagon road again and commence wading. Reached a small prairie (five miles from Mud Prairie) at sundown and camped.

April 10.—Started at sunrise ; reached Mud Prairie at 9 a. m. ; hung our snow-shoes on the willows, the snow having disappeared entirely, and pushed on for the last crossing, which we reached at 12 ; were ferried over by an Indian to whom I paid a shirt. Arrived at the Mission at 12½ p. m., made arrangement to start in the morning ; after mature reflection we have come to the conclusion to hire a canoe here to take us as far as the St. Joseph's river. Our horses were on said river, and by sending after them would detain us at least three days. The Indians would not let us have a canoe as we did not have cash to pay them for the use of it.

In the morning Rev. Father Gazzoli concluded to let us take his boat. We went to work and hewed at some oars after a fashion, and got everything ready to start at 12 m. The father furnished thirty pounds of flour, for which I was to leave an order at McClenchy's for the same amount, for one of the fathers from Portland.

April 11.—Started from the Mission at 12 m., reached the ferry on the Cœur d'Alène river at 3 p. m., where we found a large encampment of Indians; they said to us that it was impossible for us to go through with a boat, as the river was not open. Placing little or no confidence in their statement, I pushed on down the river and reached a small prairie at 9 p. m., and camped on the left bank about two miles from the lake.

April 12.—Started at 6 a. m.; soon made the lake and pulled across some four miles; was obliged to put to shore, as the ice was not yet broken; made our boat secure and were forced to shoulder our packs again. Two Indians from the Mission, who travelled with us, assisted us in packing our things, one of those Indians that had charge of our horses on the St. Joseph's river. We struck nearly a straight line across to the ferry; were lucky in finding our horses ere reaching the St. Joseph's; caught up four; leaving two in charge of the same Indian, pushed on to the ferry, which point we reached at sundown; crossed, moved down three miles, and camped.

April 13.—Started at sunrise and soon reached the bridge, which looks firm as yet, and appears to be capable to resist the pressure of high water unless molested by the Indians, who are numerous here at present; pushed on, on the wagon road; good travelling; made Col. Steptoe's council ground at 3 p. m.; continued on thirteen miles further, where we found an excellent camp, and camped at sundown. Distance travelled 41 miles.

April 14.—Started at 6 a. m.; good travelling; road hard and dry; reached Pelouse creek at 12 m.; rested the animals; pushed on to the south branch, which we reached at 4 p. m. This had swollen considerably; crossed and moved on to the first crossing of the main stream; was successful in finding a good ford 500 yards above the wagon road; crossed and camped. Distance travelled 40 miles.

April 15.—Started at 7 a. m.; struck a small Indian trail, which we followed down on the north side of the Pelouse; good travelling, but somewhat rocky. Struck the Colville wagon road at 11 a. m., which we followed to Snake river; crossed two men and four horses, and camped at 5 p. m.

April 16.—Started at 5 a. m.; pushed on rapidly, determined to make Walla-Walla if possible; reached the Touchet at 12 m.; rested our horses one hour, and then pushed on; made Walla-Walla at sundown; delivered the mail at the quartermaster's office to Captain Kirkham; turned in four horses, saddles and bridles.

April 17.—Drew fifteen days' rations, commencing on the 20th of April, 1860, and ending on the 4th of May, 1860.

April 18.—Started out to the Touchette to the quartermaster's herd; received four horses, and returned on the 19th; arrived in the garrison at 1 p. m.; the balance of the day was passed in preparing to start in the morning. Captain Kirkham gave us two pack-saddles.

April 20.—Received of the quartermaster a portion of the mail, and went over to town, where Mr. Charles N. Mullan had the balance of it put up for us. Started from town at 10½ a. m.; reached the Touchet at 3½ p. m., and camped. This river had fallen three feet two inches since the 16th.

April 21.—Started at 7 a. m.; reached the ferry on Snake river at 3½ p m.; crossed two men and four horses; pushed on, on the Colville wagon road, and camped at sundown on the Pelouse, about nine miles from Snake river.

April 22.—Started at 6 a. m.; travelled very slow, horses scarcely went out of a walk; the ground was literally covered with rock; camped at sundown at the upper end of a lake, (suppose it to be the Spokane.)

April 23.—Started at 7 a. m.; at one we left the wagon road and took a straight line for the fishery on Spokane river; travelled until sundown; camped sixteen miles from the fishery.

April 24.—Started at 6 a. m.; reached Spokane river at 12 m.; found one family here, packing up to leave for the camas ground; got ferried across by one of the Indians; swam our horses; paid the Indians six plugs of tobacco rather than to travel sixty miles further, which we would have had to do, had we not crossed here. Here also we discovered that Lieutenant Harker and Dr. Hammond, with a party of soldiers, had passed a few days previous on their way to the Clark's Fork to establish a supply depot there for the north boundary commissioners. We pushed on twelve miles, and camped.

April 25.—Started at 5½ a. m.; reached Antoine Plantes at 9 a. m.; stopped and rested our horses. Antoine had gone to Colville with Mr. Ogden a few days previous; we were also informed that Major Owen passed yesterday, taking Spokane Garry as his guide. Travelling on we camped at sundown; excellent grass for our animals.

April 26.—Started at 7 a m.; reached Clark's Fork at 4 p. m Here Lieutenant Harker and his party are encamped; he said to me that he had a wagon road cut from Spokane prairie to his camp on the Clark's Fork, and said he expected a party to join him in a few days from Walla-Walla; they were to travel on the Fort Walla-Walla and Fort Benton military wagon road as far as mile-post 150, thence to the Spokane river, and soon as they would join him he would start and cut a high water trail from Clark's Fork to Coutanie lake. The lieutenant having no conveniences here by which we could cross, we were necessarily compelled to hire an Indian and his canoe; he crossed us in four trips. Paid him nine plugs of tobacco. Camped on the north bank.

April 27.—Started at 6 a. m.; pushed on along the lake; at 5 p. m. came to a large stream; here we met two men from Hell-Gate who had just crossed; they said that their horses had to swim in crossing. We took our packs off and carried them in front of us on our riding horses; got all our things over dry; packed up again and moved on three miles, and camped.

April 28.—Started at 7 a. m.; travelled very slowly; the trail somewhat rocky; no grass for our animals; moved on until dark, and camped on the bank of the river.

April 29.—Started at sunrise; travelling until noon, same as yesterday; from that time until night the trail began to improve, as also the grass. Camped at sundown.

April 30.—Started at 6½ a. m.; reached Vermillion creek at 11 a. m.; pushed on seven miles further and camped. Rained all night.

May 1.—Started at sunrise; reached Thompson's prairie at 10½ a. m., and the crossing of Thompson's river at 11 a. m.; pushed on to that well-known point called "Bad Rock," which we reached at 4 p m. Camped.

May 2.—Started at 7 a. m.; passed that rocky point all right; moved on at a tolerably quick gate; made Horse Plains at 10 a. m. No Indians here at present. Moved to the upper end of the prairie to look for wood enough to build us a raft, as it is our only hope now. At the upper end, however, we were successful in finding driftwood enough to build us a concern on which to cross, securing it well with rope; piled all our things on board; pushed her in; went across, but considerably down stream; got all our things across in good order. Camped; excellent grass for our horses; swam the horses.

May 3.—Started at sunrise, and followed a small Indian trail, which led over a very steep mountain and down to the Bitter Root river. Some deep patches of snow on the north side of this mountain yet. Reached Bitter Root river at 11½ a. m.; moved up the river to the ferry, which point we reached at 4 p. m., and camped.

May 4.—Started at 8 a. m.; reached Cantonment Jordan at 3½ p. m.; delivered the mail; turned in to Lieutenant Lyon four horses, two riding-saddles, two pack-saddles, all complete.

P. E. TOOHILL,
Expressman to Military Road Expedition.

SIR: I have forwarded, from time to time, reports of the expressman carrying the mail, (for, as you may know, there is no mail-route in this vicinity,) in order that the department, as well as myself, should be truthfully informed of the condition of the road at all seasons of the year, winter as well as summer, during seasons of freshets as during seasons of fording. These have enabled and aided me in forming not only correct views regarding the road, but have also pointed out the contingencies to which it is liable, and the improvements yet needed before it can become a permanent line of travel.

I have had these reports made, also, in order that we should lose no opportunity in collecting any and every information regarding the route and the road, and in order that we may be enabled to speak from well-founded data, and not, as heretofore, on this, as on many other routes and lines, upon guesswork, conjecture, and surmise.

I am, sir, truly and respectfully, your obedient servant,

JOHN MULLAN,
First Lieut., in charge of Military Road Expedition.

Captain A. A. HUMPHREYS,
U. S. Top. Eng'rs, in charge of Office Exp. and Surveys,
War Department.

Q.

CAMP ON TOUCHET RIVER, *W. T.*, *September* 26, 1860.

SIR: Having been charged by you with the duty of examining the valley of the Hell-Gate river and the approaches to Mullan's Pass of the Rocky mountains, together with the succeeding part of the line from thence to Fort Benton, all along the line of the wagon road, as to their practicability for a railroad line, I beg leave to present the following as the results of my observations:

From Brown's, at the commencement of the Hell-Gate road, to Observatory creek, a distance of eighteen miles, there is no difficulty in locating a road, as the ground is of a very easy character, being a gently rolling plateau elevated ten or twelve feet above the Bitter Root and Hell-Gate rivers. Above this, again, there is another plateau about one hundred feet high at the foot of the mountains, a point of which (over which the wagon road passes,) projects out close to the junction of the two rivers.

This plateau was thought at first to present a shorter and more favorable line, but subsequent examinations proved it to be cut up by deep ravines, and to have a very irregular surface, so that very high grades and deep cuts and fills would be necessary for the whole distance. The lower plateau, next to the water, is intersected by three small creeks, which would require bridges of from thirty to fifty feet. Grades would not exceed twenty-five feet, except at the above-mentioned projecting point, where a grade of fifty feet would be necessary on either side. From thence to Observatory creek (four miles) there would be a descending grade of twenty feet per mile. At the creek an embankment of ten feet high on either side, and five hundred feet in length altogether, with a bridge fifty feet in length, would be necessary. From Observatory creek to the Big Blackfoot river, distance seven miles, the line would run on a plateau for about one mile and a half, to a point where it breaks off, and the low river bottom commences, which lasts for over half a mile. At the end of this flat is a clay bluff one hundred feet high and five hundred feet long, with a slope of 45° abutting directly on the river. It ends at a small creek, the left bank of which is a plateau fifty feet high, which extends to the Big Blackfoot river. An embankment with a grade of fifty feet would be necessary across the flat, and a cut through the bluff and twenty feet culvert across the creek, with one hundred and fifty feet embankment. This would reach the plateau on the opposite side, and from thence to the Big Blackfoot the grades and work would be very light, as the surface of the plateau is very regular. There is a small creek to cross on the plateau. On the left bank of the river there is also a plateau corresponding nearly in height to that on the right bank. The best point for a bridge is not far from the foot of the hills, and about half a mile up the river from its mouth, where the plateaus are nearly of the same height and come down directly to the river. The bridge would be about one hundred yards long and elevated about forty feet above high-water mark. The Big Blackfoot has a very swift current, with large boulders of rock in its bed. It rises in high water about five to seven feet.

From the nature of the ground, passing from one large stream to another with a projecting plateau between, and from the sinuous course of the Hell-Gate after entering the defile near Observatory creek, it is inevitable that there should be in the above distance a large amount of curvature. Short tangents, however, can be obtained, and none of the curves would be less than half a mile radius.

Within the next twenty-eight miles after leaving the Big Blackfoot the wagon road crosses the Hell-Gate river eleven times, besides several sloughs and one large stream called the South Branch, and, six miles before the last crossing, passes over the neck of a spur projecting into the river from the right bank, called the "Beaver Dam." All of the crossings except the last one are avoided in high water by trails which pass along the side hill, always on the right bank. In high water the banks of the river are not more than from two to four feet above the water. Between each of the crossings there is a flat, of which those on the right bank are generally the longest and most elevated, some of them having plateaus of from ten to twenty feet high. The best location for a railroad, in my opinion, would be along these hill-sides, where the trails are, so as to avoid the numerous crossings and sloughs. About one-third would be in rock cutting and loose rock, when retaining walls would be necessary. The whole distance to be thus worked would be about eight miles. Some projecting points would require to be cut through, making short tunnels of from fifty to two hundred feet. These would be culverts and embankments over some three or four small creeks. Some of the work, particularly on the first and last of the trails, would be very heavy, but occur at points where there are several crossings and sloughs. These hill-sides are well formed to admit curves, none of which would be sharp, and the general grade would not exceed forty feet. A tunnel of about three hundred yards would be necessary through the spur called the Beaver Dam.

It should be remarked here that up to the last crossing the left bank of the stream is very rocky and precipitous, and much less favorable in every respect than the right bank, particularly at the crossings.

Beyond the last crossing the right bank becomes very unfavorable, the hill-sides being very rocky and precipitous for several miles. About half a mile above a spur projects out from the left bank, which forms a rocky and narrow cañon, through which the river passes for about half a mile. Beyond this the left bank is flat and very favorable for the next ten miles. The line would pass the river at the last crossing with a bridge of about 80 yards length, and a tunnel would be required of about 700 yards through the spur. This would give a much straighter line than going round by the river, and would be very little more work, if any. The flat beyond this on the left bank is generally broad, with only one or two points where grades exceeding 40 feet would be required, and bridges over a few small sloughs and creeks, not exceeding six in number nor over 80 feet in length. The excavation and embankment would be very light, and many straight lines could be obtained; no curves short of one-half mile radius.

Eleven miles above the spur the bluffs, on both sides, come down close to the river forming a cañon, which extends up the river for some four or five miles. The left bank is very rocky and steep next the river, whilst the right bank is much more favorable with but few rocky points. The best location would be to cross the stream below the entrance of the cañon bridge, 70 yards long, and keep entirely on the right bank. At the upper end the valley opens out, and there are flats on both sides. For the next 30 miles the line would still continue on the right bank, as it would be straighter and less work than on the left, besides avoiding another crossing of the river and bridging Flint and Benetzi's creeks. The latter part of the distance would be undulating, and there would be about one and a half mile of deep grading, but not much rock. Grades not higher anywhere than 50 feet, and curves of smaller radius than quarter of a mile; a few small creeks would need bridging, not exceeding 40 feet in length. At the end of this portion of the line, and nearly at the junction of the Little Blackfoot and Deer Lodge rivers, (which form the Hell-Gate,) there is a projecting mountain spur on the right bank. This is very rocky, particularly upon the river, and is lowest where the wagon road crosses. It is about half a mile across. It might be overcome by grades of 60 feet for about three miles on either side, but on account of the ground beyond it would perhaps be better to tunnel it.

Beyond this the line would follow the course of the Little Blackfoot river. This stream is not more than 20 yards wide at any place. Its banks are generally low, and its course very winding. About two miles above the junction is the entrance of a cañon, through which the river runs for at least 10 miles. There are three points where it is very narrow with rocks on either side; between these, however, the cañon widens out considerably, and there are flats on either side of the creek. At the upper end the creek runs between perpendicular walls of rock for at least a mile. The last location, considering the irregular formation of the cañon, and to avoid many sharp curves, would be to locate the line entirely in the valley, crossing the stream as often as may be necessary. Of these crossings, I think there would be about eight. By this means short tangents could be obtained, and there would be no curves of less than quarter of a mile radius. The grades would not exceed 40 feet per mile. Beyond the cañon the valley of the river opens out, and presents no difficulty as far as the forks of the stream. The line would take the left bank for about six miles, crossing at a favorable point where the banks are high, and would then take the right bank for five miles to the forks of the stream, which are two in number, the northern and southern branches. In this distance the work generally would be light.

The northern branch is the one which would be taken by the line, "Mullan's Pass" being at the head of it. The distance to it from the forks is about seven miles. The stream is very small, not being more than 15 feet wide at its mouth; the upper part of the valley is wider than the lower portion, and the whole forms a very easy and gradual approach to the pass, which is very low and easy of passage on both sides. From the forks to the divide there would be a general grade of not more than 60 feet to the mile, most of which would be side-hill excavation without much rock, for which the right bank is the most favorable, and probably about three-quarters of a mile of embankment with an average height of 15 feet. There would be three crossings of the stream, and the line would be a very straight one.

The tunnel through the divide would not exceed a mile in length, with a descending grade of 70 feet to the opposite side, where the Big Prickly Pear, a branch of the Missouri, heads. The valley of this stream is narrow, but without sharp bends; the hills are of limestone, and would not be difficult to work. The left bank is the most favorable. The line would follow down on this side for about nine miles to Silver creek, a branch coming in from the north. There would be one cañon in this distance, where the work would be very heavy for about three-quarters of a mile. From the divide the general grade would be about 80 feet for about three miles, and then about 60 feet. The direction east, or as far as Silver creek, where it turns to the northeast. The line would then proceed up Silver creek to the head of one of its branches over to the waters of Little Prickly Pear creek. These are all in the same valley, and the divide between them is very low. This valley is, however, surrounded by high rocky hills. This portion of the line would be easy to work, with but little rock, and no grade higher than 50 feet. From the point where the line strikes the main branch of the Little Prickly Pear it would follow this stream down to the point where the wagon road leaves it within a mile and a half of the Missouri, a distance of 19 or 20 miles. This section would be the most difficult and expensive portion of the line, as the creek runs mostly through a succession of narrow and winding cañons with precipitous rock on both sides. The stream itself is very crooked, and would require a very large number of bridges, (average length, 40 feet,) and very sharp curves, as well as embankments in places to raise the line above the level of high water. The grade would be a descending one of about 45 feet until within about four miles of the wagon road, where it would take the left hand side hill in order to reach the level of the high rolling ground which extends for 16 miles between the Little Prickly Pear and Dearborn rivers.

This section is elevated about 200 feet in its highest part above the level of the two rivers, and is intersected by some five or six ravines running down to the Missouri river, making a very rough and broken country. The line would cross these ravines, and the work would be very heavy, with high grades. There would, however, be no sharp curves, and but little rock cutting At the Dearborn river there would be an embankment or trestle-works 60 feet high for about one-eighth of a mile, and bridge 120 feet long to the left bank, which is the edge of a plateau. From thence the line would cross by a gentle ravine to the valley of Beaver creek, across which there would be an embankment 15 feet high for one-quarter of a mile with bridge 50 feet across the creek, and take the left bank of the creek and its northeast branch to ts head, and cross over to Bird-tail Rock creek, down which it would follow to near its junction with Crown Butte creek. The line would then follow down this latter creek some three miles, in order to avoid a rocky spur, and would then cross it and pass to the left of the butte known as the Big Knee to Sun river ; the distance from the Dearborn would be 33 miles. The work would be nearly all in gravelly earth, except, perhaps, in crossing the divide from Beaver creek to Bird-tail Rock creek, when rock would probably be found. The line would be crooked, but the work would be generally light, and the maximum grade not exceed 50 feet per mile. The approach to Sun river is very good, and about two or three miles above the present Indian agency the bridge over the river would be about 200 feet long.

From Sun river to Fort Benton there extends a high rolling plateau. At Sun river where the road crosses it is about 100 feet high, and it afterwards rises by steps to the height of probably 500 feet. It is entirely destitute of timber, and has but little water. The line by which I passed over it would not be the railroad line, and I had no time to make any side explorations, as I was attending to the odometer line during the day, and also made the greater portion of the march by night. As far as I went down Sun river (eight miles) there exists no difficulty in the way of a railroad.

Material for the construction of a railroad can be found generally close at hand all along the line of the road, more particularly on the Hell-Gate river, and near the pass on both sides. There is an abundance of pine timber on the surrounding mountains, and stone of suitable quality for masonry, and which could be easily dressed. There is abundance of limestone at the divide, the main ridge of which seems to be entirely composed of it.

I have endeavored in the above report to represent the difficulties and amount of work on the various parts of the line as well as I am able to judge of them by observation, unassisted by instruments. The grades and amount of work wherever specified are spoken of in the maximum.

In conclusion, I would remark that whilst I regard the line of Hell-Gate, by "Mullan's Pass," as eminently practicable and easy for railroad purposes, notwithstanding a few difficult points, it would require a much more extended exploration and survey to determine the best line of location on the Missouri side. The line along the Missouri river would require to be examined, in order to find if possible an easier location than that by Little Prickly Pear creek. I had an opportunity of examining a portion of this line, from the mouth of the Little Prickly Pear to the mouth of Dearborn river, and found it much more favorable as regards grades and curves and generally in the character of ground, except near the mouth of Dearborn river, where for about three miles it was very rocky.

I am, sir, with much respect, your obedient servant,

W. W. DE LACY,
Chief Engineer.

Lieut. JOHN MULLAN, *U. S. A.*,
In charge of Military Road Expedition.

R.

CAMP AT NEW FORT WALLA-WALLA,
October 12, 1860.

SIR: The following report of an exploration for a line for a wagon road from the Cœur d'Alène Mission, *via* the outlet of the Cœur d'Alène lake, to Antoine Plante's on the Cœur d'Alène prairie, at the crossing of the Spokane river, and a reconnaissance thence, *via* what is known as Mix's road to Snake river, is respectfully submitted for your consideration.

In obedience to your verbal instructions of the 13th of September, I left the Mission on the morning of the 14th and proceeded as far as the Wolf's Lodge prairie, which is the first camp ground from the Mission on the line of the Spokane trail. The Wolf's Lodge is a prairie a half mile in length and a quarter mile in width, with a fine clear stream, fifteen feet wide and two deep, flowing through the eastern and northern portion of it. This stream is lined with willows, the prairie is covered with excellent grass, and is enclosed by open pine timber. Its distance from the Mission by the trail is about 18 miles. Here I camped, and on the morning of the 15th of September started with Bazil, my Indian guide, to examine on foot the line from the prairie to the Mission.

As it is impracticable to follow the line of the trail, leading, as it does, over high steep hills, rocky in many places, covered with fallen timber, and also making ascents and descents so steep as to endanger even a rider, the only way therefore in which a road can be had is to follow the lines of the water-courses in the valleys below the trail, and work such places in side-hill excavation as are rendered otherwise impassable for a wagon road.

After leaving the trail, at the east end of the Wolf's Lodge, where it first takes the side hill, we passed through pine and fir timber for a hundred yards, when we entered a small prairie, 200 feet in diameter, after passing which there are no more prairie openings. We followed near the foot of the hills to the north, and for the first mile and a quarter passed through pine and fir of the size usually seen in the wooded country on the mountains near the Mission, the trees being from 80 to 150 feet in height and from 8 to 20 inches in diameter. There is an old Indian trail for part of this distance. We then turned to the northward, crossed a dry water-run and ascended along the backbone of a spur for a quarter of a mile to a saddle of about 200 feet in height. The ground was strewn with fallen timber, which covered the ground in every direction, rendering our progress slow and laborious. In 900 feet more made the descent from this saddle, which is somewhat steep on the east, but will not require doubling teams. After reaching the foot of this saddle, which I will designate as No. 1, the line for the next mile and a half turns more to the north and follows up the narrow valley of a small stream, crossing from time to time as may be necessary. The bed of this stream is of hard gravel; one thousand feet of light side-hill grading will have to be done to pass through a narrow defile made by the passage of this stream through a spur of the mountain from the southwest, and for a quarter of a mile more granite boulders of from one to two feet in diameter will have to be removed in order to allow the easy passage of wagons. The line strikes the trail about a mile east of saddle No. 1, and follows its general course for half a mile, when it leaves it to the north and ascends an open ravine, narrow at the bottom and very much obstructed with fallen timber, which brings us to saddle No. 2, by a good ascent of a quarter of a mile in length and 200 feet high. The descent to the east here is steep and will have to be modified by grading, unless a natural descent be found on the spurs leading to the south. A valley is reached in the bottom of a steep ravine, along which the road will pass, and in a mile and a half ascend to a third saddle, which will be reached by a long and not very steep ascent of three hundred feet in height.

This distance of five and a half miles is difficult. The ground is covered with fallen timber, which is crossed and tumbled in every direction, rendering progress exceedingly difficult. The ascents and descents to these three saddles are steep, not sufficiently so, however, as to require double teaming. It will not be possible to find any other practicable line than the one explored, as the mountain ridges extend in a direction parallel with the course travelled, and rise to a height of from one to two thousand feet. Between the first and second saddles there are two or three streams, which come from this range, and are crossed near their mouths by the line of the road location. The ground is hard and solid, being composed of gravel and vegetable mould, and the streams are never of sufficient size to render them impassable, and in the fall are nearly dry.

Descending from the third saddle by a very good slope to the valley, 200 feet below, we reach that portion of the line which is most difficult and will cost much labor in the process of construction. In the first mile, 1,400 feet in length will have to be graded with the cross section. This brings to the first cañon, where the stream flowing from saddle No. 3 enters—a cañon which is from 25 to 50 feet in width, walled with perpendicular rock some 20 to 30 feet in height. Here, for one mile, the road will have to be cut in earth and rock excavation along the side of a spur, with a transverse slope of 40°; advantage can be taken of the formation of the ground so as to avoid any very severe work except for short distances. After passing this we can again descend to the valley of the stream, follow on its left bank for 200 yards, when we cross a small tributary from the north and come to the second cañon, similar in its character to the first; the stream winds through it with a narrow alder bottom, alternately forming the banks as the creek turns from side to side in its descent. This cañon is two and a half miles in length, and will be avoided by side-hill excavation for that distance. The work on this section will be severe, and if the rock proves to be hard when reached may have to be worked out bv blasting. It outcrops in many places, especially on the points of spurs, and has a transverse slope from 20° to 30°.

After passing the second cañon the ravine opens out by the receding of the mountain spurs, and a cedar swamp begins, which continues for three and a half miles. The spring can only determine whether it is best to locate the road through this, which has heretofore proved to be objectionable ground. At the time when I passed over it, it was good and dry. The huge cedars, from three to eight feet in diameter, towered above, and their interlaced branches hardly admitted the sun's rays. If this is found hard in spring, but little work will be required to pass through, as the trees grow sufficiently wide apart to allow the passage of a wagon conveniently. We crossed two large dry water-courses, with gravel bottoms, which were four feet below the level of the surrounding ground. A more detailed examination may show that a line may be had by keeping around the foot of the spurs, and doing occasional grading at points where required, thereby avoiding this long stretch of ground, which is soft and miry.

After leaving this cedar swamp we struck good pine timber, and clambered over and between fallen trees for a quarter of a mile, when we reached a spur with a stream at its base. Here a crossing may be made, but it would be best to grade three hundred feet around it, and then continue for nearly a mile over the same kind of ground to a second spur, around the base of which the line would be graded for one thousand feet. Here the line leaves its general southeast course and deflects towards the east for a short distance, when it again resumes its general direction. At the point of this spur we leave the main valley of the creek and follow up a small tributary, to shorten distance. I will here note a small cone, about fifty feet high, which forms a landmark, and round the western base of which the line passes. In a mile we ascended by an easy grade to a low saddle or ridge, and then descended as easily to the east, and after passing over good ground for a mile and three-quarters reached the open prairie, one and a half mile from the Mission.

I regret that I could not make this examination as thoroughly as I desired. My guide, the Indian Bazil, had recently broken his collar-bone, and the fracture was not even dressed, so that he was unable to cary our blankets and provision on his back. This work, devolved, therefore, upon myself, in addition to my other duties ; and it was not possible for me, thus encumbered, to examine in person the country in the vicinity of the line. The only place requiring further examination is the distance around the Cedar swamp of three and a half miles, at the base of the spurs.

I estimate that eighteen days' work of fifty men will be required to *cut* and *clear* the road from the Mission prairie to the Wolf's Lodge ; and for the *grading*, if the rock proves compact in the side-hill work referred to, a month, or even six weeks, would not be too large an allowance for the time which fifty men would occupy in working it. There are at least four and a quarter miles of grading required to avoid the two cañons of the creek and the other points noted in the foregoing. The distance estimated by the line travelled is seventeen miles.

Having finished this portion of the exploration I returned to the Wolf's Lodge, and on the 21st proceeded to examine the line to the outlet of the Cœur d'Alène lake. For the first two miles and three-quarters the line would follow the general direction of the trail, by taking advantage of the ravines to the south in gaining the ascent of three hundred feet to a saddle. This ascent is made in the first mile and three-quarters. A good descent can be found from this saddle to the valley of a small creek in a mile more, although it will have two or three steep grades in that distance. After crossing this stream, which flows through a valley of nearly a quarter of a mile in width, the trail winds up the steep side of a mountain, which is nearly five hundred feet high, and impracticable at that point as an ascent for a road. After gaining this elevation the ground is nearly level for the next three and a quarter miles, with the exception of three ravines, which are crossed by the trail. The first of these occurs a mile and three quarters west of the creek, and is broad and level; width three hundred feet. This will be subsequently referred to as the third ravine. The other two occur in one-half and three-quarter mile further on; are steep and narrow, coming to a line at the bottom, fifty feet below the general surface of the ground, presenting a serious obstacle to the passage of a road, without resorting to heavy bridging and trestle work. The descent to the lake's shore is made by a sharp pitch of four hundred and fifty feet.

Having heard of Antoine Plante as a man who, from his long residence in this neighborhood, was well acquainted with it, I deemed it best to proceed to his place and secure his services as guide, so as to be able to find and examine a line of which the Indians have spoken, leading direct from the Wolf's Lodge to the lake, and giving good ascents and descents, in place of the steep pitches which occur on the line of the trail. Accordingly, I proceeded to the outlet of the Cœur d'Alène lake—distant four miles from the foot of the mountain—and then followed down the Spokane river, on the line of the trail, for six miles, when we encamped and remained a day to examine the Little Falls of the Spokane, concerning which a report and sketch is herewith submitted.

Having finished this examination, I started on the 23d, and in two miles left the open pine timber through which I had travelled since leaving the Wolf's Lodge, and entered upon the Cœur d' Alène prairie, which stretched a level plain as far as the eye could reach to the northeast, towards the Pend d'Oreille. This prairie is bounded by isolated ranges of mountains, which, with their spurs timbered from base to summit, form a pleasing contrast to the prairie covered with its dry and browned grass. The Spokane river flows through and near the southern end of the prairie in a channel some thirty feet below the general level of the country. The soil is gravelly and not suited for farming purposes until you cross the Spokane, after which, as I am informed by Antoine Plante, the soil is fertile. After travelling nine miles across the prairie, we reached the foot of the mountains on the west, along which we proceeded for three miles, when we arrived at Antoine Plante's.

On the 24th I started back with Antoine to complete the examination of the line from the lake to the Wolf's Lodge. We followed the trail as far as the mouth of the Cœur d'Alène lake, (head of the Spokane river,) and then struck to the east through the open timber for four miles, until we were stopped by a lake a mile in length and half a mile wide, whose existence was not before known. This lake is enclosed on its northern, southern, and eastern sides by high, steeply-sloping, rocky mountain spurs, which forbid the passage of a road on any other line than that in the vicinity of the trail. Having satisfied ourselves of this, we followed down the western side of this lake, and in a quarter of a mile reached the shore of the Cœur d'Alène lake, where we encamped, and on the morning of the 25th proceeded to examine a ravine which leaves the trail to the south, and affords an ascent, somewhat steep in places, of a mile and a half in length to the summit of the last mountain over which the trail passes before reaching the lake. The ascent in this distance is not far from 1,000 feet, and will afford a road without any very heavy work—one, however, which will require double teaming to make the ascent. The descent on the east to the third ravine is good, but to improve it may require 300 feet of light side-hill excavation in earth. The ascent of fifty feet to the plateau east of this ravine can be made by side-hill grading for 350 feet, if a natural one cannot be found on further examination. A couple of hundred yards from the top of this plateau we again left the trail and struck the head of a ravine, down which we followed for a mile and a half to the valley of the creek, entirely avoiding the steep grade previously referred to. This ravine is cañon-like in many places, its bottom is covered with a growth of generally small fir trees, and is somewhat obstructed with fallen timber. It is only wide enough to allow the comfortable passage of a wagon in many places, say fifteen to twenty feet wide. This is the only ravine which can be taken advantage of to make the descent, and, if not used, side-hill excavation, part in rock, must be resorted to for at least a mile, to make a practicable grade. The ravine debouches a few yards from the trail to the north. After crossing the creek the ground is less difficult, and the only obstacles are those previously noted.

Antoine Plante informed me that the snow disappeared from these mountains in May, there being perhaps a little left in deep ravines protected from the sun ; its average depth during mid-winter is stated not to exceed two feet. On the Cœur d'Alène prairie there is but little or no snow, the Indians always resorting to it as a common wintering ground for their horses and cattle.

The line thus explored from the lake to the third ravine is objectionable on account of its steep grades, which can only be obviated by a mile and a third of heavy grading along the face of the mountain forming the eastern shore of the lake, to the third ravine, where it can be located for the remainder of the distance to the Wolf's Lodge by following the line of the trail as before indicated. As there will be but little cutting to be done through this open timbered region, I consider a month's work of fifty men as abundant for making a wagon road from the head of the Spokane to the Wolf's Lodge, and a more careful examination may show a line which can be rendered practicable in one-half the time.

We returned to Antoine Plante's and rested a day, before resuming our march upon the line known as Mix's road, towards Snake river.

My guide, Antoine Plante, is a man over fifty years of age. He is a half-breed, his father being a Canadian and his mother a Gros Ventre. He has resided for thirty years on this side of the Rocky mountains, has travelled all over the country up to the headwaters of the Columbia, in the British possessions, and down even into California. He has resided upon the Cœur d'Alène prairie for sixteen years, and has a wife and several children. His old house having been burned in Colonel Wright's campaign, in 1858, he has erected a new one out of hewn timber dovetailed together, with a verandah and steps in front, and window sashes neatly painted and glazed. He has a fine enclosure of about fifteen acres, which has evidently yielded a good crop. A small house was filled to the eaves with sheaves of wheat, bunches of corn with full and yellow ears were drying out in the sun, potatoes, turnips, beets, onions, &c., were peeping from out the ground, and the fat animals around and within the enclosure showed the evidences of good living. He has every preparation made for enclosing ten acres more, which he intends to sow in wheat this fall.

Just below his house is the ferry of the Spokane, where he has built a ferry boat capable of carrying a loaded wagon and its cattle. The river here is one hundred and twenty-five feet wide, and about eight feet deep. It is, however, fordable six miles above.

We crossed the river on the morning of the 27th September and resumed our march. Travelled for five miles over level prairie, and then entered open timber, which continued for one mile and a half, when we entered a prairie skirting around the spurs of the mountain range, forming the western boundary of the Cœur d'Alène lake; in two miles and a half more reached a small stream, which sinks just above the crossing of the road. Here is a good camp ground, with sufficient water for stock, and cotton-wood for fuel. Continued for one mile and a half over good level ground, when the road makes a detour to the left, passes down a ravine, then turns to the right, and continues for a mile on a level plateau, when it descends one hundred feet to the Camas Prairie creek, on the Ned-l-hwad-lk—distance twelve miles from the crossing of the Spokane. Road crosses this stream, which is now twenty feet wide, and five to twelve inches deep, but at high water its width is fifty feet; followed up its left bank through pine and fir timber for a quarter of a mile, when the road makes a steep ascent of ninety feet, which requires double teaming, (needs grading to improve it.) Road thence takes an easy plateau through strips of small pines and open timber, and rises gently towards the west. Five miles from the Ned-l-hwad-lk reached a series of marshy prairies, which are the headwaters of the Ciel-ciel-pow-vet-sin. The road winds around over good ground, avoiding marshy places, passing through strips of small pines for the next six miles, when it goes over slightly rolling prairie, and in nine miles more joins the military road one mile east of its crossing of the Ciel-ciel-pow-vet-sin. In this distance of twenty-one miles, between the Camas Prairie and Ciel-ciel-pow-vet-sin creeks, there is no water during the autumn, but there is every indication of an abundance at intervals of every few miles in the marshy bottoms, which occur frequently along the road. Timber is to be had everywhere.

Leaving the military road some two miles west of the Ciel-ciel-pow-vet-sin, we turned more to the west, following a succession of alkaline bottoms, which evidently have water in them in spring and early summer. In five miles passed a landmark, consisting of a tall pine, with a long green cone at its top. Called this the mistletoe pine. After passing this tree, followed a basaltic basin with alkaline bottoms for three miles to the Three Buttes, between which the road passes. These buttes form prominent landmarks, and from them westward we see no more pine timber. After travelling a mile and a half from the Three Buttes, descended to a valley where a road to a camp ground turns off to the west. Ascended to a plateau in a mile and three-quarters, and travelled along at the base of a prairie plateau seventy-five feet high. Passed some water amongst some cotton-wood bushes in two miles and three-quarters more. The tracks of wagons and remnants of camp fires indicate this as an old camp ground. Seventeen miles from the Ciel-ciel-pow-vet-sin is a good camp ground. A spring flows from out the base of the upper plateau, yielding sufficient water for stock. Wood consists of willows, but the supply would be exhausted in a few seasons. A mile below this point I noticed an old wagon road leading to the north, and passing into the pedrigal region, which we see extending for some eight or ten miles to the westward of the road.

Seven miles from the spring last mentioned we passed another old camping ground, and in seven miles and one-quarter more reached the Aspen camp of Colonel Wright, on his way up the Spokane, in in August, 1858. The entire line thus far is remarkably level, and affords an excellent road. No work will be required between this point and the Ciel-ciel-pow-vet-sin. In this distance of thirty-one miles, the only camping ground at present available for wagon trains is the spring fourteen miles northeast of the Aspen camp, giving a distance of seventeen miles thence to the Ciel-ciel-pow-vet-sin. In spring and summer there is, doubtless, water at the other points noted as old camp grounds.

Leaving the Aspen camp in seven miles, reached a spring, and in an eighth more a second, giving abundant water for animals, with willows and cotton-wood for fuel. Six miles further are two springs, which, if cleared out, would afford sufficient water for camping purposes. In two miles further reached a valley an eighth of a mile wide, which has a small water-course (now dry) passing down it. Followed this for four miles, where it debouches into the valley of Cow creek. Here we entered the travelled road leading from Walla-Walla to Colville. From the Aspen camp to Cow creek, 19.8 miles by odometre measurement, the road follows a series of alkaline bottoms and the valley last referred to, passing over good ground. The crossings of the dry water-course in the valley only will require a few hours work.

The Colville road, going towards Snake river, follows down the valley of Cow creek for about two miles, when it ascends to the upper plateau three hundred feet, and then continues along a valley for five miles, when it ascends a steep hill five hundred feet high; continues two miles over a ridge, and in two more reaches the Pelouse, which is a general camping ground at this point, although wood is scarce, and in the latter part of summer and fall the grass becomes eaten off for miles around. By driving across the river an abundance of grass is found close at hand

From this point to Snake river, twelve miles, the road ascends to a plateau three hundred feet high, and then continues until within a mile and a half of Snake river, when it descends by a couple of steep pitches through sandy ground to the river banks. This distance of twenty-three miles between Snake river and Cow creek is the worst portion of the road, but is continually travelled by the teams supplying Colville, Harney depot, and the northwest boundary commission.

Snake river at the crossing is four hundred yards wide, and the flying ferry, running on a three-inch wire rope, serves as a very secure and speedy communication from bank to bank.

Having thus terminated my reconnaissance, the above report is respectfully submitted by your obedient servant,

W. W. JOHNSON, *C. E., &c.*

Lieutenant JOHN MULLAN, U. S. A.,
In charge of Fort Walla-Walla and Fort Benton Mil. Road Ex.

Respectfully submitted to Captain A. A. Humphreys, United States topographical engineers, War Department.

JOHN MULLAN,
First Lieut. 2d Artillery, in charge of Military Road Exp.

CAMP AT FORT WALLA-WALLA, W. T., *October* 26, 1860.

S.

CAMP AT NEW FORT WALLA-WALLA,
October 20, 1860.

SIR: I have the honor to present the following report of an examination of the Little Falls of the Spokane, made by me on the 22d of September last, in connexion with a reconnaissance from the Cœur d'Alène Mission, *via* Antoine Plante's to Snake river, in compliance to your instructions of the 13th ultimo.

The Spokane river flows from the Cœur d'Alène lake, and is a stream of from 150 to 250 feet in width. Its present level is about 15 feet below its banks, and the water is shallow. It is fordable at the first rapids, which are some four miles below the lake, whence to the Little Falls, a distance of four miles more, the stream flows over a succession of light gravelly rapids, from one to three-quarters of a mile apart. At the Little Falls, eight miles below the lake, the river has forced its way through a mass of granite rock 90 feet in height and 900 feet long. The channel through this rock being about 60 feet wide, passing straight in its course, with perpendicular walls on each side. The left bank, at the falls, is a wall of from 80 to 90 feet high above the water, while the right is perpendicular for about 15 feet, when it breaks off and rises less precipitously to a height of from 30 to 40 feet above water level. The stream enters this channel, or cañon, by a fissure, F, (see accompanying sketch,) about eight feet wide between the rock island E and the projecting rock G. A small stream forces its way through an opening blocked up by the boulders 1, 2, 3, 4. These streams unite below the island E. The basin above the falls is about 140 feet wide and from 25 to 30 feet deep, forming a reservoir, which discharges itself through the fissure F, the orifice being 8 feet wide by 10 deep. This empties into a narrow rocky channel, which is about 15 feet below the level of the basin above. The water rushes in foam through this channel, which is about 30 feet wide, and in 100 feet falls some 15 or 20 feet more over large boulders. The stream thence is confined in the cañon of 60 feet width, and flows through it with a more gentle current for about 600 feet, when it debouches into a basin about 200 feet in diameter, whence it flows over a succession of light gravelly rapids, as before, towards the Columbia.

The ordinary rise of water above the falls is 15 feet, but the highest flood-mark is some 25 feet above the present level. At high water the rocks H and I, forming the cañon of the river, become islands; for the water, after rising 8 feet, flows around on either side, having a channel 150 feet wide, now dry, on the left bank, at the southern foot of the granite rock I, and on the right bank a channel of 50 to the north of the rock H. The general cross section will show the appearance of the stream at high water.

If the rock island E, which is 60 feet long and 15 feet in height, with a mean width of 10 feet, the boulders 1, 2, 3, 4, and the low rock G, (see section A B,) be removed, the river would have as free a channel at low water as it now has after having risen five feet; and the conclusion is, that the water being thus allowed to pass off rapidly when it first rises, and the channel increasing in dimensions with the rise of the river, sufficient clearance will be had to prevent the great rise to which the river and lake are now subject, and relieve the valuable and extensive lands bordering the lower Cœur d'Alène and St. Joseph's rivers from that overflow which at present renders them impassable and unfit for agricultural or other purposes until late in the summer. The amount of land so reclaimed would be about fifty square miles, or an area of 32,000 acres, all of which is well adapted both by soil and climate to all the purposes of cultivation.

It is estimated that an appropriation of $5,000 would be sufficient to defray all the expenses of the work.

The above is respectfully submitted for your consideration by your obedient servant,

W. W. JOHNSON, *C. E., &c.*

Lieutenant JOHN MULLAN, *U. S. A.,*

In charge of Fort Walla-Walla and Fort Benton Mil. Road Exp.

The foregoing throws additional light upon the subject of the removal of the higher Spokane, with a view of improving the line of the military road *via* the valleys of the St. Joseph's and Cœur d'Alène, to which I have several times called the attention of the department, being sanguine in the belief that not only will the road be much improved, but a valuable tract of 32,000 acres of beautiful soil thus reclaimed. If it be within the legitimate province of the department to take any action in the matter, I trust the subject may be brought before Congress in connexion with our present operations.

JOHN MULLAN,
1st Lieutenant 2d Artillery.

CAMP AT WALLA-WALLA, W. T., *October* 31, 1860.

T.

The following is a statement of the character and amount of work performed on the military road from Fort Walla-Walla to Fort Benton, under direction of Lieutenant John Mullan, U. S. A., during the summer and fall of 1859.

From Fort Walla-Walla to Snake river, at the mouth of the Tukanon, distance forty-nine miles, the road passes over the natural surface of the ground. Soil is clayey. High rolling hills occur on the line of the road, and their summits are gained by easy ascents. Two streams occur in this distance which will require bridging, viz: the Dry and Touchet creeks. Length of bridges required, seventy-five feet each.

Snake river is crossed by a ferry-boat. Width of stream is a quarter of a mile. The ascent from Snake river to the high plateau table land is steep, but cannot be avoided. It is easily overcome by doubling teams.

From Snake river to the St. Joseph's river, near its entrance to the Cœur d'Alène lake, a distance of one hundred and nineteen and five-tenths miles, the road passes for sixty-three miles over soil of the same character as before mentioned, through a gently rolling prairie region, bordered by high plateaus. The remainder of this distance, fifty-six and five-tenths miles, is through open pine timber, with gravelly soil. The road, on nearing the St. Joseph's, was cut through timber for some six miles, and was graded, for a hundred feet, through limestone rock.

The descent to the valley of the St. Joseph's is made along the backbone of a spur of the mountain range, and a substantial bridge, sixty feet long, was thrown across a slough or branch of the river. A flatboat, capable of transporting a loaded wagon and four yoke of cattle, was built, and remains at the crossing of this river for the use of the subsequent trains. A low, level flat exists for about six miles along the St. Joseph's river, across which wagons cannot pass until towards the middle of July. Means for obviating this will be hereafter pointed out. From the St. Joseph's to the Cœur d'Alène river, distance ten and eight-tenths miles, the road lies over a gravelly rolling plateau, covered with open timber, through which the road was cut and cleared, to obtain the most direct line. The rise to this plateau is easy, and the descent to the valley of the Cœur d'Alène river was accomplished by side-hill grading, some ten thousand cubic yards of excavation being done at this point. The flats bordering the Cœur d'Alène river are followed up for ten miles, and are subject to the same overflow as those of the St. Joseph's. This is caused by the backing up of the Cœur d'Alène lake, its outlet, the Spokane river, being too small to admit of the rapid passage of the large body of water created by the melting snows of the mountains, which, from a very large extent of country, are drained into the lake by these two rivers. It is deemed practicable to enlarge this opening at the falls of the Spokane river by blasting, so that the waters of the lake and rivers will be lowered some three feet, and a very beneficial result will also ensue to the Indian tribes, by allowing the salmon to enter these waters and add to the stores of their now scanty and precarious subsistence. The amount of work required for this purpose cannot now be stated, but it is deemed to be practicable at no very great outlay.

From the first crossing of the Cœur d'Alène river, nine miles below the Mission to the camp of party, fifty-eight miles east of the Mission, the road has been constructed through a heavily-timbered region up the valley of the Cœur d'Alène river, across the divide of the Cœur d'Alène mountains, and down the valley of the St. Regis Borgia river, to within thirteen miles of the Bitter Root river. The soil is gravelly. Nearly one hundred crossings have been made of both rivers, which must be bridged in order to render the road passable at all seasons of the year. High water lasts from the beginning of May to the middle or end of June, during which time travel is suspended. All of these crossings can be bridged without difficulty. The banks are high, and timber is abundant as well as convenient. The average length of these bridges will be fifty feet each. The bridge over the Bitter Root river, which is indispensable, will be two hundred and fifty feet long, and should be constructed at an early day. A flatboat is already provided for this crossing. The amount of excavation done on the divide of the Cœur d'Alène mountains was about forty thousand cubic yards, and the entire amount done on the road during the season is about sixty thousand cubic yards. There are some two miles of road to be relocated and altered, whereby two crossings of the river will be avoided and a better road made.

The width of the road as cut and cleared through timber is from fifteen to twenty feet, and its length constructed this season, from Fort Walla-Walla to Cantonment Jordan, is two hundred and fifty-eight miles.

Respectfully submitted.

W. W. JOHNSON, *C. E.*,
Fort Walla-Walla and Fort Benton Mil. Wagon Road Exp.

www.ingramcontent.com/pod-product-compliance
Lightning Source LLC
LaVergne TN
LVHW011238110826
845150LV00006B/1670

* 9 7 8 1 4 2 5 5 1 3 9 0 0 *